Water Resources Engineering

Water Resources Engineering

Editor

Petre Kozel

scitus
academics

Water Resources Engineering
Edited by **Petre Kozel**

ISBN: 978-1-68117-178-4
Library of Congress Control Number: 2015950364

SCITUS Academics LLC,
www.scitusacademics.com
Box No. 4766, 616 Corporate Way,
Suite 2, Valley Cottage,
NY 10989

Printed in United States of America on Acid Free Paper

Table of Content

Preface

Water resources engineering is the quantitative study of the hydrologic cycle the distribution and circulation of water linking the earth's atmosphere, land and oceans. Surface runoff is measured as the difference between precipitation and abstractions, such as infiltration (which replenishes groundwater flow), surface storage and evaporation. Applications include the management of the urban water supply, the design of urban storm-sewer systems, and flood forecasting. Hydraulic engineering consists of the application of fluid mechanics to water flowing in an isolated environment (pipe, pump) or in an open channel (river, lake, and ocean). Civil engineers are primarily concerned with open channel flow, which is governed by the interdependent interaction between the water and the channel.

The purposes of this book, includes both water use and water excess management. The fundamental water resources engineering processes are the hydrologic processes and the hydraulic processes. The hydrologic processes include rainfall, evaporation, infiltration, rainfall-runoff and routing, all of which can be further subdivided into other processes. Knowledge of the hydrologic and hydraulic processes is extended to the design and analysis aspects.

Water resources development has had a long history basically beginning when humans changed from being hunters and food gatherers to developing of agriculture and settlements. This book builds on that knowledge to present state-of-the-an concepts and practices in water resources engineering.

CHAPTER 1

Robust Decision Making and Info-Gap Decision Theory for Water Resource System Planning

Evgenii S. Matrosov, Ashley M. Woods, Julien J. Harou

Department of Civil, Environmental and Geomatic Engineering, University College London, Chadwick Building, Gower Street, London WC1E 6BT, UK

SUMMARY

Stationarity assumptions of linked human–water systems are frequently invalid given the difficult-to-predict changes affecting such systems. In this case water planning occurs under conditions of deep or severe uncertainty, where the statistical distributions of future conditions and events are poorly known. In such situations predictive system simulation models are typically run under different scenarios to evaluate the performance of future plans under different conditions. Given that there are many possible plans and many possible futures, which simulations will lead to the best designs? Robust Decision Making (RDM) and Info-Gap Decision Theory (IGDT) provide a structured approach to planning complex systems under such uncertainty. Both RDM and IGDT make repeated use of trusted simulation models to evaluate different plans under different future conditions. Both methods seek to identify robust rather than optimal decisions, where a robust decision works satisfactorily over a broad range of possible futures. IGDT efficiently charts system performance with robustness and opportuneness plots summarising system performance for different plans under the most dire and favourable sets of future conditions. RDM samples a wider range of dire, benign and opportune futures and offers a holistic assessment of the performance of different options. RDM also identifies through 'scenario discovery' which combinations of uncertain future stresses lead to system vulnerabilities. In our study we apply both frameworks to a water resource system planning problem: London's water supply system expansion in the Thames basin, UK. The methods help identify which out of 20 proposed water supply infrastructure portfolios is the most robust given severely uncertain future hydrological inflows, water demands and energy prices. Multiple criteria of system performance are considered: service reliability, storage susceptibility, capital and operating cost, energy use and environmental flows. Initially the two decision frameworks lead to different recommendations. We show the methods are complementary and can be beneficially used together to better understand results and reveal how the particulars of each method can skew results towards particular future plans.

INTRODUCTION

Water resource systems are sensitive to climate and population changes, making supply infrastructure planning difficult. Planning models grapple with the inherent uncertainty of future conditions when the statistical distributions of future conditions are unknown or not trusted. Under such 'Knightian' uncertainty (Knight, 1921) uncertainty is unquantifiable and the most likely realisation of the future is unknown. In such situations methods that rely on traditional Bayesian decision analysis to characterise uncertainty using probability theory may not be appropriate (Groves and Lempert, 2007). In such situations of 'deep' or 'severe' uncertainty, recent research has argued it is more appropriate to strive for robustness (Ben-Haim, 2001, Dessai and Hulme, 2007, Lempert et al., 2006 and Lempert and Collins, 2007) rather than optimality. A 'robust' system performs satisfactorily, or satisfices (Simon, 1959) performance criteria, over a wide range of uncertain futures rather than performing optimally over the historical period or a few scenarios.

Robust Decision Making (RDM) (Lempert and Popper, 2003) and Info-Gap Decision Theory (IGDT) (Ben-Haim, 2001) are two decision making frameworks that seek robustness. Both use trusted simulation models to consider a wide spectrum of plausible futures each with different input parameters to represent uncertainty.

Both approaches have been applied to water management. Groves and Lempert (2007)use RDM to identify vulnerabilities of the California Department of Water Resources' California Water Plan (CWP). Lempert and Groves (2010) apply RDM to identify climate change vulnerabilities of the Inland Empire Utilities Agency's 2005 Integrated Water Resource Plan and to develop a more robust plan including adaptive strategies. Hipel and Ben-Haim (1999) use IGDT to represent different sources of hydrological uncertainty. IGDT was also used within a water resources – timber production management problem in Australia to test the robustness of planting strategies to uncertainties in wildfire return periods and the knock-on impacts on municipal water availability (McCarthy and Lindenmayer, 2007). The sensitivity of UK flood management decisions to uncertainties in flood inundation models was investigated with IGDT (Hine and Hall, 2010). To our knowledge this paper constitutes the first IGDT application to water supply planning.

In a recent comparison of the two approaches, Hall et al. (2011) highlight the strengths of IGDT and RDM for robust system planning. In that application they find both tools come to similar conclusions to a climate change problem but provide different insights about the performance and vulnerabilities of the analysed strategies. They describe IGDT as a tool comparing candidate strategies' performance under a wide range of plausible futures (robustness) and their potential for rewards (opportuneness) under favourable future conditions. RDM identifies under which combination of future conditions a particular strategy becomes vulnerable to failure through 'scenario discovery' (Bryant and Lempert, 2010). Identifying different failure conditions provides scenarios to test plans and devise new strategies. IGDT by contrast, provides the facility to simultaneously compare the robustness and opportuneness of multiple strategies but does not quantify their vulnerabilities.

Since both methods are broadly similar (Hall et al., 2011), most analysts will use either one. Our study shows this can lead to the analysis missing important information that could have been uncovered if both frameworks had been used. Joint application reveals complementary information that would not be available if only one method were implemented.

This paper provides the first joint application of RDM and IGDT to a water resource system planning problem: identifying robust water supply portfolios for London and the Thames Basin for estimated future conditions in 2035. We critically assess how the two methods help formulate robust plans and how joint use increases understanding and insights.

The paper first reviews RDM and IGDT theory (Section 2). Section 3 describes the Thames basin and Section 4 introduces the system's capacity expansion problem and future uncertainties. Sections 5 and 6 describe the RDM and IGDT applications and Section 7 discusses the lessons learned from each application,and thelimitations and the benefits of joint implementation. Section 8 concludes the paper.

BACKGROUND ON SELECTED PLANNING METHODS

Robust Decision Making (RDM)

RDM is a planning framework designed to help decision makers formulate robust plans for the future under conditions of Knightian, or 'deep' uncertainty (Lempert and Collins, 2007). RDM represents such uncertainty by considering system performance under a wide range of futures. RDM favours the concept of robustness over optimality and assumes that a strategy that is able to satisfice minimum performance criteria over a wide range of plausible futures is preferable.

RDM helps decision makers develop new strategies that are more robust than those initially considered (Hall et al., 2011). In RDM, the initial preferred plan is evaluated under a wide spectrum of plausible futures. RDM characterises the vulnerabilities of the initial strategy using a process known as *scenario discovery* (Bryant and Lempert, 2010,Groves and Lempert, 2007 and Lempert et al., 2006) which rigorously identifies sets of future conditions, or 'scenarios', where the system under the preferred plan would be under most stress. Decision makers use this information to improve the candidate strategy generating new strategies that hedge against those conditions which most frequently cause system failures. The new strategies can be resubmitted into the RDM framework and the process repeated iteratively until a suitably robust strategy is found.

An RDM analysis begins with the selection of one or more candidate strategies. Sometimes current policy can be selected as the candidate strategy or it can be one or more proposed future plans. Other times it can be chosen through a traditional utility or regret analysis (Lempert and Groves, 2010). The second step characterises the vulnerabilities of the candidate strategy by identifying under which combination of uncertain conditions it fails to meet performance criteria. To identify the vulnerabilities a trusted simulation model calculates system performance criteria for different combinations of input conditions representing a wide spectrum of plausible future states. Each run is evaluated as a success if it is able to satisfice minimum performance criteria or a failure if it is not. Performance criteria are either absolute thresholds where a strategy fails if its performance crosses a certain threshold such as a cost limit, or they are relative, where the performance of the candidate strategy is compared to the performance of an ideal strategy in the same state of the world; for example using regret (deviation from optimality) (Savage, 1954). Analysts then use cluster-finding algorithms to identify under which combinations of future conditions a particular strategy becomes vulnerable (failure clusters). Clusters are regions of the solution space bounded by one or more dimensions that have a relatively high density of failure points compared to the density in the whole of the solution space.

In the final step planners propose improvements to address the vulnerabilities uncovered in the previous step developing alternative strategies or discarding the strategy altogether if its vulnerabilities are unacceptable. The process returns to the second step to analyse an improved strategy or to the first step for a new strategy, and the process is repeated until planners agree on a 'robust' strategy. After this process, planners produce trade-off summaries which are used to compare alternative strategies.

Info-Gap Decision Theory (IGDT)

Info-Gap Decision Theory (IGDT) (Ben-Haim, 2001) is a non-probabilistic technique seeking to maximise robustness of a decision given minimum performance requirements. IGDT characterises the uncertainty of future system stresses as a group of nested sets defined by the parameter $\tilde{u}$, a best estimate of a future parameter, u. The deviation between u and $\tilde{u}$ is scaled by h, which represents an increment or 'horizon' (Info-Gap parlance) of uncertainty. This can be parameterised such that, $h : h \geqslant 0$. The Info-Gap uncertainty model is constructed as a nested set based on $\tilde{u}$ and h such that $U(h, \tilde{u})$. Fig. 1 shows a schematic of an Info-Gap uncertainty model showing the scaling of each interval (horizon) of uncertainty from a best estimate.

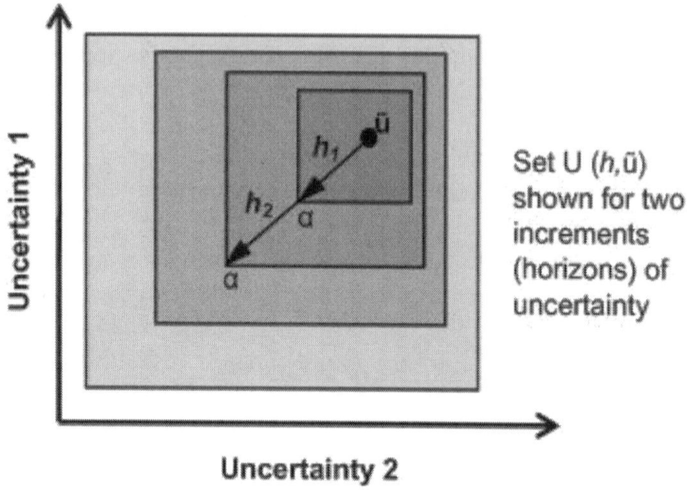

Figure 1. Schematic of an Info-Gap uncertainty model showing the scaling (h) of each interval (horizon) of uncertainty (α) from a best estimate ($\tilde{u}$). Note that the uncertainty model can be asymmetric to better represent the uncertainty surrounding the best estimate.

In the context of water resource planning, a number of management options q_i are available. Simulating each option quantifies minimum performance Π, or a performance windfall (reward) R, for different horizons of uncertainty α, such that $\Pi(q_i, h)$ or $R(q_i, h)$.

For management option q_i there will be a range of performances or rewards for each horizon of uncertainty specified by $U(h, \tilde{u})$. There will be minimum and maximum levels of performance for each h, which are referred to as *robustness* and *opportuneness* respectively in IGDT (Ben-Haim, 2001).

IGDT identifies management options that perform acceptably well under a wide range of conditions; seeking robustness rather than optimality in a process known as *robust-satisficing* (Ben-Haim, 2010). Acceptably well is defined as some minimum level of system performance that must be achieved, Π_c.

The resultant *robustness* $\hat{h}$ is a measure of the maximum amount of uncertainty α (the maximum horizon of uncertainty) that can be tolerated whilst still ensuring a level of performance Π that is greater than a critical level Π_c.

$$\hat{h}(q_i, \Pi) = max\left\{h : \min_{u \in U(\alpha, \tilde{u})} \Pi(q_i, u) \geqslant \Pi_c\right\} \quad (1)$$

As the requirement for performance increases; robustness decreases such that robustness is zero, $\hat{h}(q_i, u) = 0$, when the maximum possible level of performance $\Pi(q_i, u)$ is required. The robustness function (Eq. (1)) charts the performance of a management option as incrementally increasing uncertainty produces system inputs that result in lower system performance $\Pi(q_i, \tilde{u}) \leqslant \Pi(q_i, u)$. Uncertain inputs may also result in a level of performance Π for a given management option that produces a performance windfall Π_w, which gives rise to the IGDT 'opportuness function':

$$\hat{\beta}(q_i, \Pi) = min\left\{\alpha : \max_{u \in U(\alpha, \tilde{u})} R(q_i, u) \geqslant \Pi_w\right\} \quad (2)$$

Info-Gap assumes users can identify a best estimate of the unknown parameter from which to start the uncertainty analysis; in our case the central estimate was used because we did not have evidence to suggest we should not use it. In cases of 'severe' uncertainty this estimate may in fact be a poor estimate of the real value. If so, because Info-Gap only considers a subset of the full uncertainty space around the best estimate, it could potentially ignore areas in which the real parameter value could fall. Info-Gap seeks local robustness around the choice of the best estimate instead of global robustness in the whole of the uncertainty space.

THE THAMES BASIN

The Thames river basin covers 16,133 km^2 and has 13 million inhabitants. Mean annual precipitation is 690 mm, with mean annual long-term effective precipitation at 280 mm (Merrett, 2007). Two private water companies manage the majority of municipal water supply: Thames Water Utilities Limited (TWUL) and Veolia Three Valleys Water (VTVW). The

River Thames provides over 50% of TWUL's supply and satisfies 70% of London's demand (Jones, 1983).

Raw water from the Thames is stored in ten raised reservoirs between Slough and Teddington Weir. In the Lee valley thirteen storage reservoirs are fed from abstractions on the River Lee. The two reservoir groups are connected via a bulk transfer from the Thames Reservoirs along the Thames-Lee Tunnel (Halrow Ltd., 2010). The Thames basin has two existing surface water – groundwater conjunctive-use schemes in addition to normal groundwater abstraction. The North London Artificial Recharge Scheme (NLARS) recharges excess treated water to an aquifer in the Lee valley for later use during dry periods and the West Berkshire Ground Water Scheme (WBGW) augments supply during droughts. TWUL began operation of a desalination plant along a tidal stretch of the Thames in 2009. The Thames basin and water resource system is displayed in Fig. 2.

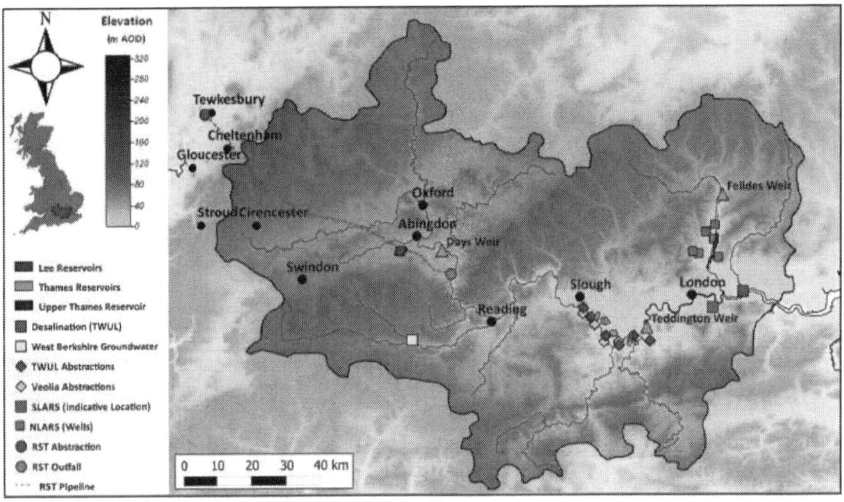

Figure 2. The Thames basin showing major urban centres, and current and possible future water supply infrastructure. Adapted from Matrosov et al. (2011).

We use a Thames Basin water resource system model built with IRAS-2010 (Matrosov et al., 2011), an open-source rule-based water resource system simulator. The model includes 28 nodes (junctions, reservoirs, aquifers, treatment plants) and 33 links (rivers, canals, transfers, etc.) and runs on a weekly time-step. IRAS-2010 tracks flows and storages of every node and link in the system and registers system performance at every time-step. At the end of the simulation performance is summarised using performance metrics. Time series of naturalised gauged flows from the National River Flow Archive (NRFA) provide inflows into the system. One aggregated surface storage node, the London Aggregate Storage (LAS), represents the combined surface storage of all the reservoirs in the basin. Abstraction to surface storage on the Thames is controlled downstream by environmental minimum flows (EMF) at Teddington Weir. The EMF varies between 800 and 300 Ml/day and is controlled by the Lower Thames Control Diagram (LTCD) (Fig. 3). The LTCD also determines when storage levels in LAS trigger the four levels of water use restrictions. For details on the model refer to Matrosov et al. (2011).

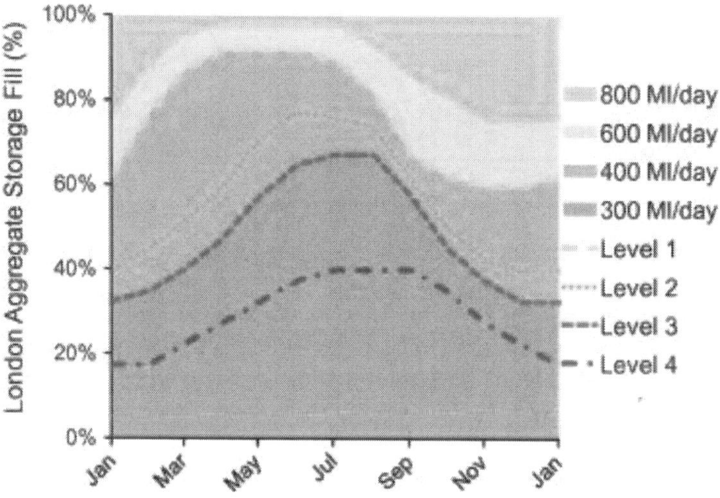

Figure 3. The lower Thames control diagram showing the four levels of minimum environmental flows at Teddington and demand restriction thresholds as a function of the day in the year and storage in the LAS. Adapted from Matrosov et al. (2011).

PROBLEM FORMULATION

The Thames basin has experienced six major droughts in the last 90 years (Marsh et al., 2007). London's population is expected to increase by 16% by 2028 (Office for National Statistics, 2008). Population growth and the likelihood that climate change will reduce summer precipitation in south-east England (Murphy et al., 2009) increases the threat of water scarcity. Water supply infrastructure expansions will form part of the solution to reduce the region's vulnerability to drought. In this case study we apply the RDM and IGDT planning frameworks to an infrastructure planning problem for 2035.

Infrastructure Portfolios (IPs)

Thames Water (2010) describes 37 plausible future water supply options including multiple variants of the same option. Of these we selected seven options (Table 1), leading to 19 possible infrastructure portfolios (IPs) (unique combinations of supply options). Selection criteria included the public availability of data and the expected impact of each option on system performance. Adding the current arrangement led to the 20 possible IPs, or 'options', considered in this case study (Table 2).

Table 1. The seven possible future options (see in Fig. 2).

ID	Name	Description	Capacity
UTR75	Upper Thames Reservoir	High capacity reservoir to supply Swindon and Oxford (SWOX) and London. Daily maximum supply set to 157.5Ml/d[a].	75 M m^3
UTR100	Upper Thames Reservoir	High capacity reservoir to supply Swindon and Oxford (SWOX) and London. Daily maximum supply set to 202Ml/d[a].	100 M m^3
UTR150	Upper Thames Reservoir	High capacity reservoir to supply Swindon and Oxford (SWOX) and London. Daily maximum supply set to 315Ml/d[a].	150 M m^3
RST	River Severn Transfer	High capacity bulk transfer of freshwater from the River Severn to the River Thames[a].	315 Ml/d
DESAL80	London Desalination	Reverse osmosis desalination plant	80 Ml/d
DESAL140	London Desalination	Reverse osmosis desalination plant.	140 Ml/d
SLARS	South London Artificial Recharge Scheme	Artificial groundwater storage and recovery scheme (South London)[a] (Fig. 2).	22 Ml/d

Table 2. The 20 water supply 'infrastructure portfolios' (IPs), alternatively called 'options' or 'plans' evaluated in the case study.

Infrastructure portfolio	Future supply options
1	DESAL80
2	DESAL80, SLARS
3	DESAL140
4	DESAL140, SLARS
5	UTR75, DESAL80
6	UTR75. DESAL80, SLARS
7	UTR75, DESAL140
8	UTR75, DESAL140, SLARS
9	UTR100, DESAL80
10	UTR100. DESAL80, SLARS
11	UTR100, DESAL140
12	UTR100, DESAL140, SLARS
13	UTR150, DESAL80
14	UTR150. DESAL80, SLARS
15	UTR150, DESAL140
16	UTR150, DESAL140, SLARS
17	RST, DESAL80
18	RST, DESAL80, SLARS
19	RST, DESAL140
20	RST, DESAL140, SLARS

We investigate the Upper Thames Reservoir (UTR) with capacities of 75, 100 or 150 million m^3 (M m^3) (UTR75, UTR100, UTR150), the River Severn Transfer (RST) and the South London Artificial Recharge Scheme (SLARS). We also investigate the operation of the desalination plant at 80 Ml/day (DESAL80) and 140 Ml/day (DESAL140).

The proposed UTR is filled from River Thames abstractions during periods of high flows. The UTR would provide continual supply to the Swindon and Oxfordshire Water Resource Zone (SWOX WRZ), with additional releases into the River Thames during drought conditions for re-abstraction downstream for London. The River Severn Transfer (RST) would import water from the River Severn to the west of the Thames basin (Fig. 2). When activated during drought conditions the RST would import an amount equivalent to the release of UTR150 (Thames Water, 2010).

SLARS pumps from the confined chalk aquifer in South London (Jones et al., 2005) and functions analogously to NLARS. For details on the function and release capacities of the UTR, RST and SLARS please refer to Thames Water (2010).

All infrastructure additions incur capital costs except for the desalination plant which is already built to operate to a maximum capacity of 150 Ml/day. Variable and fixed operating costs are considered for each proposed option.

Variable operating costs are calculated by multiplying flow from supply nodes by the unit energy consumption, and unit energy cost. Energy consumption results from desalination, pumping groundwater, to overcome elevation gain and friction loses in the RST, and pumping from the River Thames to the UTR and from the Thames Estuary to the desalination plant. We assume a pumping efficiency of 75% (OFWAT, 2010) and a consumption of 2 kW h/m^3 for desalination of brackish water (Raluy et al., 2005a, Raluy et al., 2005b and Stokes and Horvath, 2006) (Table 3).

Table 3. Costs and release rates of the management options.

Management option	Release rate (ML/day)	Capital costs (M£)	Energy use (kW h/m^3)	Fixed operating costs (k£/yr)
UTR75	133.5 to Thames 24 to SWOX[a]	536.9[a]	0.11[b]	360[a]
UTR100	178 to Thames 24 to SWOX[a]	725.5[a]	0.11[b]	370[a]
UTR150	267 to Thames 48 to SWOX[a]	821.6[a]	0.11[b]	380 [a]
RST	267 to Thames 48 to SWOX[a]	579.6[b]	1.08[b],[c]	7,000[c]
SLARS	22[a]	2.3[b]	0.23[b]	53[a]
DESAL80	80	–	2.15[b],[d]	830[a]
DESAL140	140	–	2.15[b],[d]	930[a]

[a](Thames Water, 2010).
[b]Potential energy equation with 75% pump efficiency.
[c](OFWAT, 2010).
[d](Raluy et al., 2005a and Raluy et al., 2005b).

Uncertainties

Uncertainties considered in this study include natural hydrological variation, climate change perturbation of natural hydrology, projected London water demand and future energy prices up to a 2035 planning horizon.

Natural Hydrological Variability

Significant flow variability has been observed on the River Thames since records began in the 1880s, including several periods of prolonged drought (Cole and Marsh, 2006 and Marsh et al., 2007). To consider natural hydrological variability each IRAS-2010 Thames simulation is run over the 85-year historical period (1920–2005) assuming any one of these hydrological years could occur in 2035.

Climate Change Perturbation

Climate change uncertainty is represented using monthly climate change perturbation factors that are multiplied by historical river flow time series to estimate future flows. Perturbation factors based on UKCP09 probabilistic climate change projections (Murphy et al., 2009) were used to create sets of twelve monthly flow factors (Christierson et al., 2012) valid from 2020–2035. Each set of monthly flow factors represents one hydrological future.

The UKCP09 climate projections were developed by the UK Met office and are based on "perturbed physics" and multimodel ensembles produced by a variant of the HadCM3 climate model (Johns et al., 2003) and the results of 13 other Global Climate Models (GCMs). Murphy et al. (2007) provide a description of multimodel and perturbed physics ensembles. The projections are based on the B1, A1B, A1F1 emissions found in the IPCC's (Intergovernmental Panel on Climate Change) Special Report on Emissions Scenarios (SRES) (Nakicenovic and Swart, 2000). Please consult Murphy et al. (2009)for a detailed description of the UKCP09 projections.

The Thames catchment was simulated using the Probability Distributed Model (PDM) (Moore, 1985) and the Catchmod (Wilby et al., 1994) rainfall run-off models using Latin Hypercube Sampling of the UKCP09 projections for the A1B scenario. A large number of modelled river flow times-series was generated representing the future and historical time periods to take into account uncertainty in hydrological parameters using the Generalised Likelihood Uncertainty Estimation (GLUE) methodology (Beven, 2006 and Beven and Binley, 1992). Models that closely reproduced the historical time series were used to generate a distribution of monthly flow factors. Please seeChristierson et al. (2012) for a detailed description of the flow perturbation factors.

An ensemble of 100 equiprobable flow factor sets was used. The flow factor sets were classified by their percentage perturbation effect on winter (November–April) and summer (March–October) historical flows.

For the IGDT application, the central estimate (median) flow factor set is used as the best estimate. The IGDT uncertainty model creates sets of monthly perturbation factors that diverge from the median flow factor set at structured intervals defined by the IGDT uncertainty model.

London Water Demand

London mean annual water demand estimates for the time horizon were obtained from TWUL's stochastic water demand forecasting tool (Thames Water, 2010). The demand projections for 2035 were fitted to a gamma distribution with shape 311 and scale 6.9.

For the RDM application, nine levels of London water demand were obtained from deciles of the gamma distribution. Including the demand values at the 0.01 and 0.99 percentile of the gamma distribution raised the total demand values considered to 11.

For IGDT an uncertainty model is built around this best estimate of demand $Y|x = 0.5$ (2144.5 Ml/day) with the increments of uncertainty derived conditionally from the underlying gamma distribution such that: $x = 0.525, 0.550, 0.575, \ldots, 0.975$. The same sample intervals were used to sample the left-side of the gamma distribution.

Energy Prices

A uniform distribution of price estimates (£0.09–£0.22/kW h) was assumed based on current energy price (£0.09) (Eurostat, 2011) assuming prices will increase (DECC, 2010). For the RDM application the uniform distribution was divided into 13 equiprobable intervals whose boundaries were used to produce 14 energy costs. For IGDT a best estimate of 13 p/kW h was used to construct the uncertainty model bounded by the upper and lower estimates described above.

Uncertainties for both RDM and IGDT are summarised in Table 4.

Table 4. Uncertain system parameters for strategic water resource forecasting in the Thames Basin for the 2020–2035 planning horizon.

Parameter	RDM	IGDT
Hydrological variability	Historical flows[a]	Same as for RDM
Climate change perturbation	100 monthly flow perturbation sets valid for 2020–2035[a] (Monte-Carlo sampling)	A single central estimate climate change flow factor set[b] with 40 intervals of uncertainty either side of the central estimate
Water demand	11 water demand levels obtained using LHC sampling of deciles of a gamma distribution + 2 extreme values from distribution forecasts [b]	A single central estimate of demand with 40 conditional intervals of uncertainty scaled by the gamma distribution either side of the central estimate[b]
Energy prices	14 energy prices [c] obtained using LHC sampling of 13-quantiles of a uniform distribution	A single central estimate with 40 horizons of uncertainty either side of the central estimate[b]

[a]National River Flow Archive (NRFA).
[b](Thames Water, 2010).
[c](Eurostat, 2011).

Performance Criteria

IGDT and RDM require performance criteria to classify each simulation as a failure or success. Success or failure is based on five satisficing metrics: reliability of water supply service, reservoir storage susceptibility, environmental performance, energy consumption and total costs (capital and operating).

The service reliability criterion is based on the frequency of water use restrictions imposed in the Thames basin over the full simulation. The restrictions considered are levels 2 and 3 from the LTCD (Fig. 3) which correspond to sprinkler and non-essential use bans respectively. Average service reliability is calculated by assessing the number of weeks (W_{fail}) each restriction level (i) was imposed in the simulation over the whole 85-year simulation (W_{sim}):

$$R_i = \left(1 - \frac{W_{fail,i}}{W_{sim}}\right) \tag{3}$$

Thresholds for L2 and L3 failures were calculated from the current maximum frequency of supply restrictions TWUL will impose; 1 in 10 years for sprinkler bans (level 2–L2) and 1 in 20 years for non-essential use bans (level 3–L3). For L2 restrictions we use the 9 month (40 week) hosepipe ban imposed by TWUL in 2007 (BBC, 2007) as a representative L2 restriction duration. An allowable frequency of L2 restrictions of 1 in 10 years translates to 332 weeks of allowed L2 restrictions over the 85-year simulation. L3 restrictions can be imposed no more frequently than 1 in 20 years (half that of level 2) and we use a representative failure duration of 4.5 months in line with historical durations of non-essential use bans in southeast England (WaterBriefing, 2006). Following Eq. (3), a reliability of 0.925 for restriction L2 (sprinkler ban) is used for this study and a reliability of 0.973 is used for restrictions at L3.

A storage susceptibility metric is defined as the lowest storage level reached by LAS. Its emergency storage of 22.5% of capacity (45 M m^3) is the threshold below which failure occurs due to pressure-related distribution problems in the network (Cookson and Weston, 2008).

Environmental performance is calculated downstream of the last abstraction on the Thames at Teddington Weir. A failure occurs whenever the flow at this node drops below 800 Ml/day, the minimum environmental flow during normal conditions (Fig. 3). The duration and severity of these failures are quantified using the Shortage Index (*SI*) adapted from Fredrich (1975):

$$SI = \frac{100}{N_T} \sum_{t=1}^{N_T} \left(\frac{WS_t}{WD_t} \right)^2 \tag{4}$$

where WS_t is the flow shortage at week t, WD_t is the weekly flow minimum (800 Ml/day × 7 days) and N_T is the total weeks in the simulation. Higher SI values signal worse performance.

The energy consumption criterion is the total cumulative energy consumed by infrastructure during each simulation run. The total cost criterion combines capital and cumulative operating costs of each simulation. Capital costs, fixed operating costs and energy consumption figures for infrastructure options are provided in Table 3.

Because of a lack of explicit performance requirements for the environmental, energy and cost criteria we consider the worst 15% of performances to be failures.

In our RDM and IGDT implementations, if a simulation fails in at least one of the performance criteria outlined above, that particular simulation is considered a failure.

RDM IMPLEMENTATION

We perform one iteration of the RDM framework. From among the twenty possible infrastructure portfolios (IPs) we select one as the candidate strategy. We then identify its vulnerability scenarios and determine which IP would perform better inside these scenarios to consider possible tradeoffs between IPs. Ways to address weaknesses in the IPs are considered.

We use the enumeration method to sample 20 infrastructure portfolios (Table 2), 101 climate scenarios (100 equiprobable hydrological realisations and the historical record), 11 levels of estimated London water demand and 14 values of energy costs described in Section 4.2 leading to 311,080 simulations.

Choosing a Candidate Strategy using Regret Analysis

For RDM one of the twenty IPs must be selected as the initial candidate strategy. We perform a regret-based ranking (Savage, 1954) of the proposed IPs to identify which performs best over the simulated set of plausible futures.

Regret, or deviation from an ideal performance of criteria c, is the difference in the performance P of the best performing strategy (s') and that of the strategy in question, s, for the same input parameters, j.

$$R_c(s,j) = |P_c(s',j) - P_c(s,j)| \quad (5)$$

For cost, energy and environmental performance; where low values are better, an absolute value is required. In addition to considering regret for each criteria, we also consider multi-criteria regret, where the regrets are standardised over a 0 to 1 interval and weighted equally.

Fig. 4 shows box and whisker plots of normalised regret for the five performance criteria and the multi-criteria aggregate regret. Relative performance is assessed on three ordering statistics: the lower quartile (lower end of the box), median (red line inside the box) and the upper quartile (upper end of the box) value of regret for each IP. We use median regret to select the candidate IP.

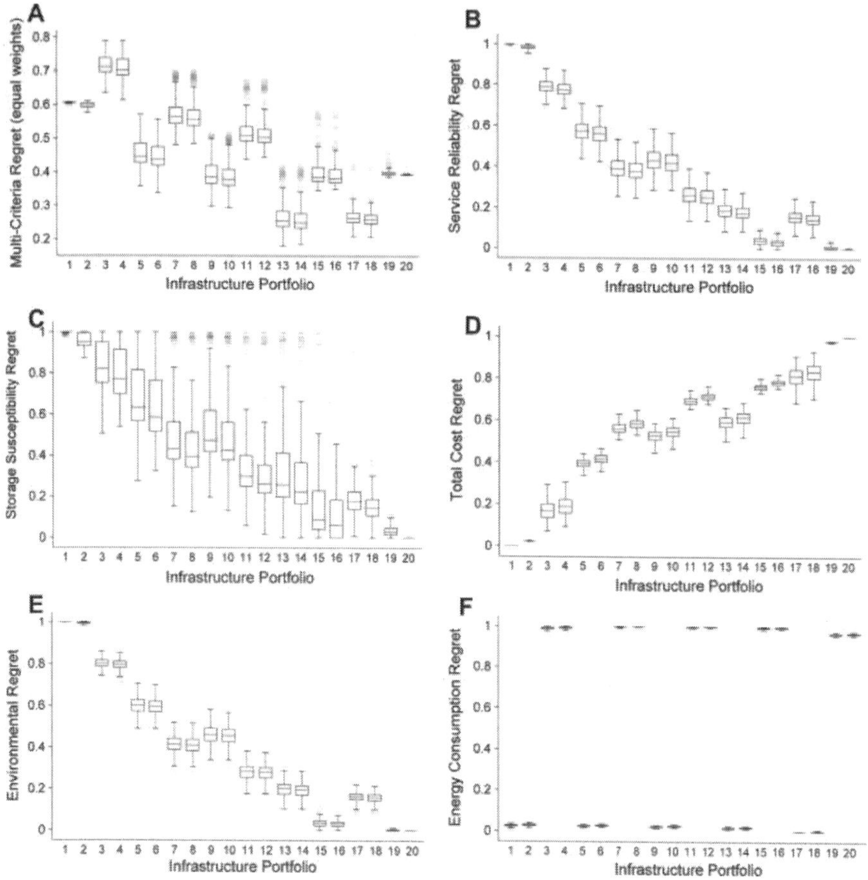

Figure 4. (a–f) Box and whisker plots of regret in each of the five performance criteria and an aggregated regret calculated using equal weighting of all criteria. The median fractional regret is represented by the red line in each box. The top and bottom of the boxes are the upper and lower quartiles respectively while the whiskers represent the regret value within 1.5 times the interquartile range from the upper and lower quartiles. Points represent outliers.

Fig. 4f (Energy consumption regret), the easiest to interpret, shows that all IPs that include DESAL140 have median energy regret close to 1 and IPs with DESAL80 have near-zero regret. Desalination is twice as energy intensive as RST and eclipses the pumping energy requirements (Table 3).

IP 17, which includes the RST, has the lowest median regret (median regret is 0) because the RST perennially provides out-of-Thames-basin flow (unlike the UTR which frequently empties during severe droughts) and thus relies least on desalination.

Fig. 4b, c and e show regret for service reliability, storage susceptibility and environmental performance. IPs with more and larger infrastructure perform better with these criteria as they can supply more water during droughts. IP 1, the baseline IP, has the highest median regret followed by IP 2 which includes SLARS. SLARS has little effect on median regret of IPs in the service and environmental performance criteria. Increasing desalination output to 140 Ml/day strongly reduces median regret in all IPs as does the UTR. Larger UTR capacities improve regret. In all three criteria the increase in regret from running desalination at 80 Ml/day rather than 140 Ml/day outweighs the amelioration in median regret that the UTR100 provides over the UTR75. In all three criteria, IP 20, which includes the RST, DESAL140 and SLARS, has the lowest median regret. IP 19 performs nearly as well followed by IPs 16 and 15 which replace the RST with the UTR150. IPs with less and smaller infrastructure (e.g. IPs 1 and 2) perform better in the cost criterion (Fig. 4d). SLARS results in a small increase in cost regret as does the higher desalination output due to high energy use. The RST incurs large capital and operating costs leading to the highest regret values. IP 20, the most costly portfolio in every modelled future, has regret values approaching 1.

Fig. 4a shows equally weighted multi-criteria regret. All options with DESAL140 have higher regret than those with DESAL80 while regret decreases with higher UTR volumes. IPs with RST and UTR 150 have similar regret values. IPs 13, 14, 17 and 18 have the lowest regret values. Table 5 summarises the lower quartile, median and upper quartile regret of the four best performing IPs.

Table 5. Upper quartile, median and lower quartiles of regret for IPs 13, 14, 17 and 18.

	IP 13	IP 14	IP 17	IP 18
Upper quartile	0.284	0.276	0.278	0.272
Median	0.254	0.248	0.262	0.258
Lower quartile	0.236	0.232	0.249	0.246

In this case study we chose the option with the lowest median value of regret, IP 14, to take forward to the next RDM phase.

Characterising Vulnerabilities of the Candidate Strategy

Having identified infrastructure portfolio 14 (UTR150. DESAL80, SLARS) as our candidate strategy, RDM's scenario discovery stage identifies under which combinations of uncertain parameters IP 14 is vulnerable to failure. Following Bryant and Lempert (2010) we use a modified version of the Patient Rule Induction Method (PRIM) (Friedman and Fisher, 1999) to characterise the vulnerabilities of IP 14. Vulnerabilities are defined as combinations of input conditions that produce high levels of failures.

PRIM is an interactive statistical cluster-finding algorithm that finds one or more low-dimensional boxes, or 'scenarios' in a hyper-dimensional space where the density of interesting points inside each box is higher than the space outside the box. PRIM seeks to maximise the density, coverage and interpretability of boxes. Density is defined as the total number of interesting points (in this case failure points) over the total points inside the box whilst coverage is the total number of interesting points inside the box compared to the total in the entire space. A trade-off or balance between coverage and density must be subjectively made by the analyst. Interpretability is how easily decision makers can interpret each box as a coherent 'scenario'. Interpretability is approximated by the number of dimensions that define each scenario.

We apply a modified version of PRIM using the "Scenario Discovery Toolkit"[1] in the publically available statistical computing environment R which implements the PRIM code with a variety of useful features and visualisations (Bryant and Lempert, 2010). The Scenario Discovery Toolkit produces density vs. coverage trade-off plots. Each point on the plot represents a unique scenario described by one or more dimensions, and each dimension is represented by one uncertain system parameter. The analyst selects the box with an appropriate density to coverage ratio. Once a box or 'failure scenario' has been identified, the failure points inside the box are removed from the solution space and a new plot can be generated in order to find additional boxes. This process is repeated until the decision maker is satisfied by the boxes identified.

Each member of the solution space for IP 14 is classified as success or failure using only four of the five performance criteria: service reliability, storage susceptibility, environmental performance and total energy consumed. Total cost was not used because all IP 14 runs have the same capital and fixed operating costs and only variable operating costs differ and are directly proportional to the energy consumption. This resulted in a failure density of 54% with the majority of failed simulations resulting from failure in the service reliability criterion.

Two scenarios were identified using PRIM covering 78% of all failure points with a density of 93% inside the two scenarios. In total the scenarios include 44% of all simulations. Fig. 5a and b shows the density vs. coverage trade-off plots generated by PRIM (called ‘peeling trajectories’ in PRIM jargon). The colours of the points represent each dimension being constricted; a change in colour represents the constriction of a new uncertainty dimension. The circled points represent the boxes chosen as having the appropriate coverage and density to describe scenarios. Choosing an alternative point in Fig. 5a would result in PRIM generating an alternate subsequent trade-off curve (Fig. 5b) as choosing an alternative scenario would result in different points remaining in the solution set.

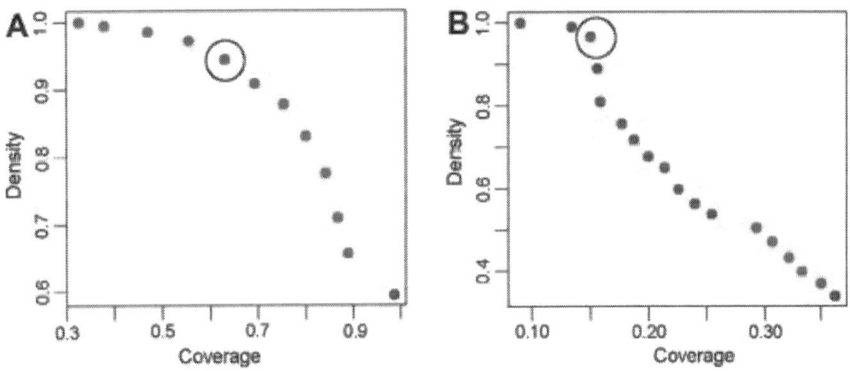

Figure 5. Density vs. coverage trade-off plots produced by the scenario discovery toolkit. Colours represent different dimensions. Circled points were chosen as scenarios. (For interpretation of the references to colour in this figure legend, the reader is referred to the web version of this article.)

The dimensions and boundaries of our chosen vulnerability clusters are outlined in Table 6. Table 7 provides the density and coverage of each box.

Table 6. Dimensions and boundaries of the two vulnerability scenarios identified using PRIM for infrastructure portfolio 14.

Scenario/dimension	1	2
Fractional change in summer hydrology	<0.92	<0.94
Fractional change in winter hydrology	–	<0.87
London demand (Ml/day)	>2098	>1933

Table 7. Density and coverage of the two vulnerability scenarios for infrastructure portfolio 14.

Scenario	Density	Coverage
1	0.94	0.63
2	0.97	0.15
Total	0.95	0.78

Scenario 1 describes futures where summer river flows are more than 8% lower than historical (1920–2005) and London's mean annual demand is greater than 2098 Ml/day. In this scenario dry summers cause IP 14 to fail, even with moderate water demands. The high coverage (63%) of this scenario shows IP 14 is vulnerable to dry summers combined with moderate to high demand levels. Change in winter hydrology is not included in this scenario because changes in winter hydrology do not substantially affect this IP. In futures where summer flows are low, the system suffers from a lack of storage and cannot buffer the summer–winter disparity in flow, even with the desalination plant providing 80 Ml/day of additional supply during droughts.

Scenario 2 describes a scenario where London's mean annual demand ranges from low (in the interval between the first and second deciles) to high and includes dry summers and extremely dry winters (6% and 13% drier than historical respectively). This failure scenario shows that IP 14 performs poorly under dry climates no matter how low demand is.

The price of energy was not found to contribute to the vulnerability of IP 14. The density of both scenarios is high (94% and 97% for scenario 1 and scenario 2 respectively) suggesting that these scenarios are exceptionally dangerous for IP 14.

Initial Findings and Recalculating Regret

Overall, moderate to high London demands combined with moderately dry summers cause IP 14 to fail. Low future demands when confronted with dry summers and extremely dry winters also lead to frequent system failure.

To check if any other IPs perform better under these vulnerability scenarios, regret is recalculated for all IPs for futures that fall within the range of conditions dictated by scenarios 1 and 2. We also examine the regret for all futures outside of the two scenarios. Regret plots for both scenarios are provided in the Supplementary material. IP 18 has the lowest median regret inside the scenarios whilst IP 14 has the lowest median regret outside each scenario. Therefore, there is a trade-off between IP 14 and IP 18. Because of the similarity between the scenarios and this single trade-off, we combine scenarios 1 and 2 into a single vulnerability scenario and recalculate multi-criteria regret for all futures inside and outside of the conditions posed by scenarios 1 and 2 (Fig. 6a and b).

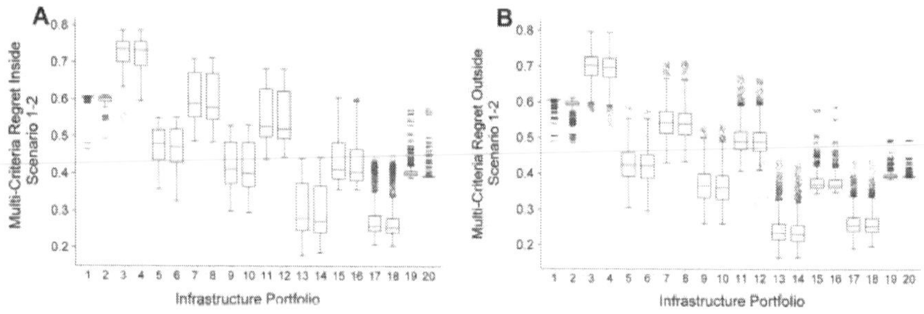

Figure 6. (a) Multi-criteria regret of all modelled futures in both vulnerability scenarios and (b) regret of all modelled futures outside the two vulnerability scenarios.

IPs 13, 14, 17 and 18 are still the best performing IPs for futures contained within the two vulnerability scenarios. Table 8 and Table 9 summarise the lower quartile, median and upper quartile regret values of the four best performing IPs. For futures inside the two scenarios, regret is higher for all four IPs. However, the increase in the range of regret for values not considered outliers is considerably greater for IP 13 and 14. IP 18 has the lowest upper quartile and median value of regret in scenarios 1–2. Compared to the increase in range of regret for IP 14, the range in regret increases minimally (median regret remains the same) indicating that IP

18's relative performance remains stable inside and outside the vulnerability scenarios. For futures outside the vulnerability scenarios, IP 14 has the lowest regret values for each of the ordering statistics.

Table 8. Upper quartile, median and lower quartiles of regret for IPs 13, 14 and 17 for futures comprised in vulnerability scenarios 1–2.

	IP 13	IP 14	IP 17	IP 18
Upper quartile	0.374	0.366	0.288	0.28
Median	0.279	0.271	0.26	0.254
Lower quartile	0.248	0.241	0.246	0.241

Table 9. Upper quartile, median and lower quartiles of regret for IPs 13, 14, and 17 for futures outside vulnerability scenarios 1–2.

	IP 13	IP 14	IP 17	IP 18
Upper quartile	0.26	0.256	0.279	0.277
Median	0.234	0.231	0.256	0.254
Lower quartile	0.215	0.213	0.24	0.238

Fig. 7 shows the median regret of each of the twenty IPs. IP 18 has the lowest median regret inside scenarios 1–2 while IP 14 has the lowest median regret in all futures outside the scenarios. Relative to the regret of every other IP however, the difference in median regret between the IP 14 and 18 is small.

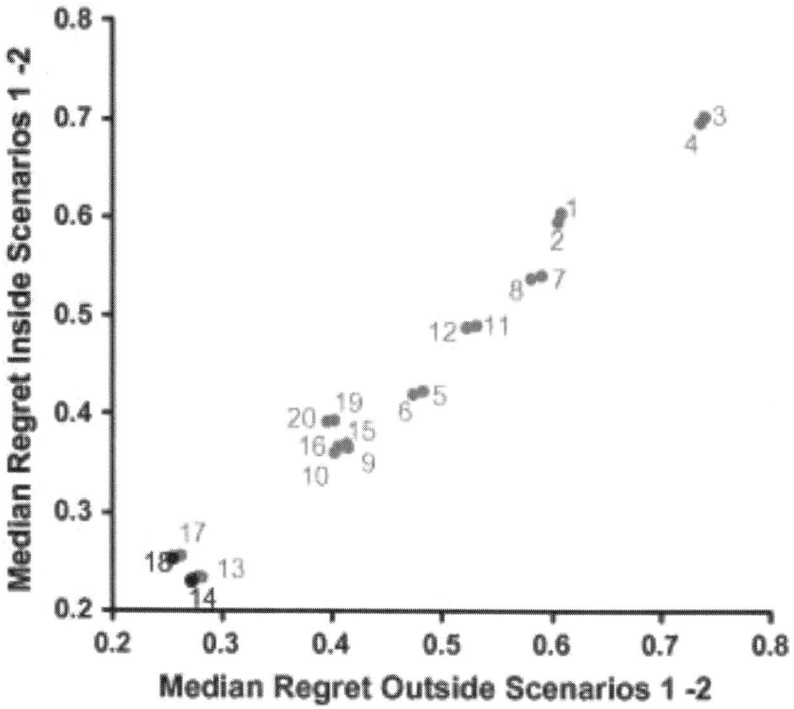

Figure 7. Median multi criteria regret of each of the twenty policies inside and outside scenarios 1–2 IP 18 has the lowest median regret inside the vulnerability scenarios while IP 14 has the lowest regret outside the scenarios.

IP 14 and IP 18 are the same except IP 14 includes the UTR150 and IP 18 includes the RST. Both the UTR150 and the RST provide the same amount of water to the Thames at the beginning of droughts. The UTR however has maximum storage and is refilled from the Thames which cannot occur during droughts whilst the RST in our model continuously provides water from a neighbouring wetter region. The uncertainty in the reliability of the RST supply (e.g. possible correlation of droughts in the Thames and Severn regions) was not considered in this study. We used a supply estimate for the transfer based on estimates of what it could provide under historical drought conditions (Thames Water, 2010).

The RST is costlier and more energy intensive, but has better performance in the storage susceptibility, service and environmental performance criteria. IP 14 however is vulnerable because it lacks the storage necessary to maintain service during extended droughts. RDM suggests that these two strategies are likely to be the most robust into the future.

Section 7.1 discusses the merits and usefulness of our RDM implementation and its recommendations.

INFO-GAP ANALYSIS

The formulation of a fractional-error Info-Gap uncertainty model for climate change perturbation, demand and energy cost is used to build robustness and opportuneness functions based on our system performance requirements.

Taking the convention u_1 = climate change perturbation, u_2 = water demand and u_3 = energy cost; κ_l and κ_r scale and, σ_r and σ_l bound the right (upper) and left (lower) intervals of the Info-Gap model respectively. The 3-dimensional Info-Gap Uncertainty model becomes:

$$U(h, \tilde{u}) = \{u : \max[\sigma_l, (1 - \kappa_l h)\tilde{u}_{i,j}] lesu_{i,j} \leqslant \min[\sigma_r, (1 + \kappa_r h)\tilde{u}_{i,j}]\} h \geqslant 0, \quad i = 1, 2, 3, \quad j = 1, 1 - 12 \tag{6}$$

where κ_l = [0.005, 0.0125, 0.1] is the scaling factors for the left hand side of the Info-Gap model for u_i; κ_r = [0.005, 0.0125, 0.25] is the scaling factors for the right hand side of the Info-Gap model for u_i; σ_l = [1.2, 1914.81, 9] is the lower boundaries of the Info-Gap model for u_i; and σ_r = [0.8, 2391.88, 22] is the upper boundaries of the Info-Gap model for u_i.

Best estimates of 2144.5 Ml/day and 13 p/kW h were used for u_2 and u_3 respectively. The best estimate for u_1 is defined by the median flow factor set from the 100 sets generated from UKCP09 climate models.

The use of scaling factors for the best estimate of the flow factors results in a compression or extension of the seasonal range of flows that is prescribed by the best estimate. When the scaling factor is less than 1 as in the case of the robustness function the seasonal range is compressed and it is extended when the scaling factor is greater than one as is the case of the opportuness function.

Robustness Function

The robustness function (Eq. (7)) from the 3D Info-Gap uncertainty model (Eq. (6)) uses the critical system performance criteria $\Pi_c(Res_o, Cost_t, Energy_p, Env_p, Rel_{L2,L3})$ to assess the performance of each management option q as a success or a failure:

$$\hat{h}(\Pi_c(Res_o, Cost_t, Energy_p, Env_p, Rel_{L2,L3}), q)$$
$$= max\{h\left(\min_{u_i \in u(h)} \Pi_{Res_o, Cost_t, Energy_p Env_p, Rel_{L2,L3}}(u_{1,2,3}, q)\right)$$
$$\geqslant \Pi_c(Res_o, Cost_t, Energy_p, Env_p, Rel_{L2,L3})\} \tag{7}$$

where Res_o is the storage susceptibility; $Cost_t$ is the total cost; $Energy_p$ is the total energy consumption; Env_p is the environmental performance; and $Rel_{L2,L3}$ is the service reliability. The relative and absolute values for each performance criteria are outlined in Section 4.3 and are the same as those implemented in the RDM analysis.

Robustness analysis of the 20 infrastructure portfolios (IPs) required 800 simulations. Our 5 performance indicators: storage susceptibility, service reliability, environmental performance, total cost and total energy consumption are used to assess a simulation as a success or a failure and the interval of uncertainty at which they fail (Table 10). Simulation results showed that portfolio IPs 1–4 failed to meet our minimum performance requirements for the best estimate scenario and IPs 5 and 6 could only cope with the best estimate scenario. Furthermore, IP 7 and 8 only maintained minimum levels of system performance for 3 increments of uncertainty, whilst IP 9 and 10 failed after 2 increments. As a result these 10 IPs were excluded from further analysis due to poor robustness. The remaining 10 IPs included 90 successful simulations. Robustness curves showing performance for IPs 11–20 at each interval of uncertainty are plotted for each performance criteria (Fig. 8).

Table 10. Robustness to uncertainty for IPs 1–20. Robustness is the maximum number of increments of uncertainty away from the central estimates of inflow, demand and cost for which the system maintained minimum performance requirements in the categories of reservoir susceptibility, service reliability, environmental performance, total cost and energy consumption.

Option	Robustness	Failure criterion
1	FAIL	Reservoir susceptibility, service reliability L2 and L3
2	FAIL	Reservoir susceptibility, service reliability L2 and L3
3	FAIL	Service reliability L2 and L3
4	FAIL	Service reliability L2
5	0	Environmental performance
6	0	Environmental performance
7	3	Environmental performance
8	3	Environmental performance
9	2	Environmental performance
10	2	Environmental performance
11	5	Environmental performance
12	6	Environmental performance
13	7	Environmental performance
14	8	Environmental performance
15	7	Energy consumption
16	7	Energy consumption
17	9	Environmental performance
18	9	Environmental performance
19	5	Total cost
20	2	Total cost

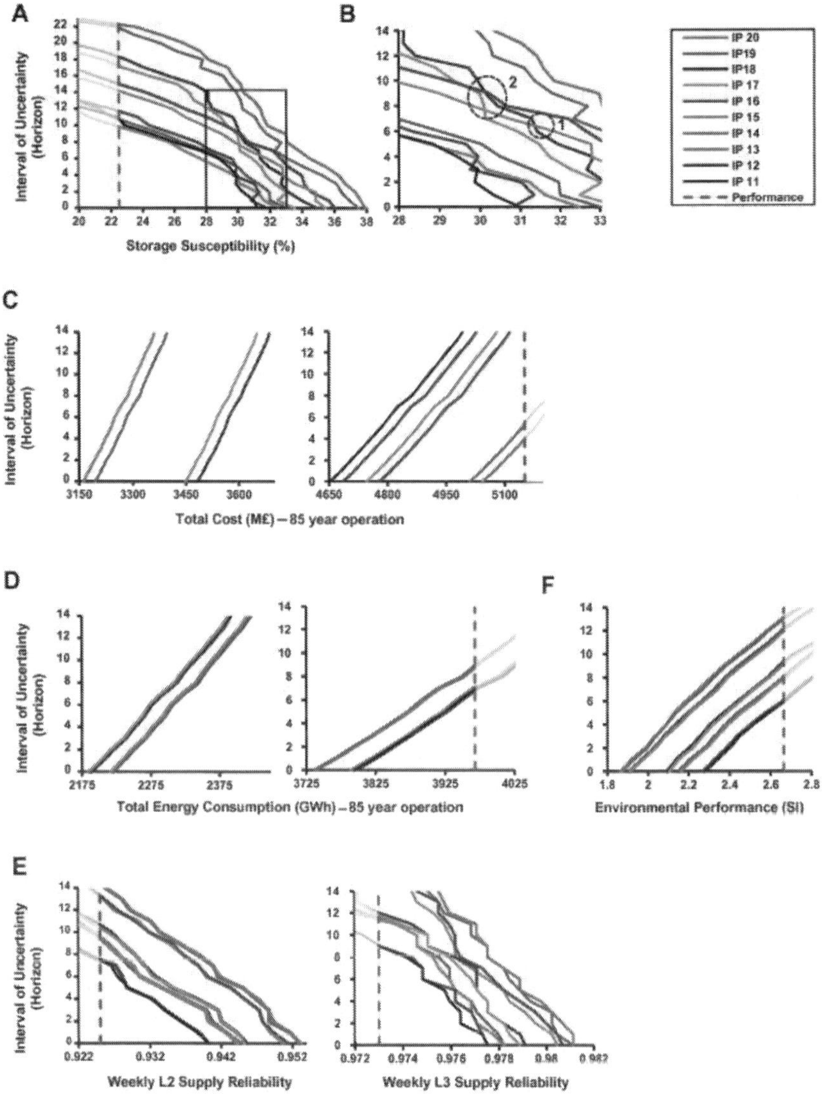

Figure 8. Robustness curves for IPs 11–20 for storage susceptibility (a and b), cost (c), energy consumption (d), service reliability (e) and environmental performance (f). Results show that for cost, energy and environmental performance there is a linear decrease in performance with increasing demand and decreasing water availability. Storage susceptibility has crossing robustness curves (1 and 2 in Fig. 8A). Where curves cross at 1, IP 18 becomes more robust to uncertainty than IP 15. At 2 the same occurs for IP 17 and IP 18; they both become more robust than IP 16.

Results show that IP 17 and 18 are the most robust, tolerating 9 increments of uncertainty from our best estimate. This translates to a future that is 5% drier than our best estimate projection, with a total potable water demand of up to 2179 Ml/day and energy cost of up to 15.25 p/kW h.

IPs 14 is the next most robust option, able to maintain acceptable levels of system performance for 8 increments of uncertainty; a future 4.5% drier than our best estimate with a total demand of up to 2175 Ml/day and energy costs of up to 15 p/kW h). IPs 13, 15 and 16 are able to cope with 7 increments of uncertainty (4% dryer, demand up to 2172 Ml/day, energy cost 14.5 p/kW h). IPs 13, 14, 17 and 18 all fail as a result of environmental performance, whereas IP 15 and 16 fail due to high energy consumption.

IPs with greater investment in supply infrastructures are shown to not be necessarily more robust to uncertainty. This is particularly true for IPs that include DESAL140, which has high energy consumption (resulting in the failures of IPs 15 and 16) and therefore also high operating costs. Combining DESAL140 with the large capital investment in RST, causes early cost failures in IPs 19 and 20 even though they are able to supply large amounts of water. Although IPs 17 and 18 also include an RST, they only include DESAL80 which significantly reduces energy consumption and costs.

Robustness curves for environmental performance, total costs and energy consumption are almost linear reflecting the relatively linear relationship these performance measures have with water availability and water demand. Steep robustness curve gradients indicate minimal performance loss as uncertainty increases, and the crossing of robustness curves marks the location where one IP becomes relatively more robust than another option.

Storage susceptibility (Fig. 8a) and service reliability (Fig. 8e) produce nonlinear robustness curves. This is because demand and level of service are a function of the time of year and the amount of storage in London reservoirs. As a result there are stages in the Info-Gap analysis were conditions get harsher and the resulting lower reservoir levels activate demand management measures sooner. Service reliability goes down as more weeks are spent in restrictions, but storage susceptibility improves when compared to the previous more benign interval of uncertainty. This is demonstrated when comparing Fig. 8b and e for IP 19, where a dramatic improvement in storage susceptibility at uncertainty interval 8 compared to interval 7 is matched by a sudden drop in L2 and L3 service reliability over the same 2 increments.

IPs 19 and 20 outperform IPs 13–18 in terms of storage susceptibility at each increment of uncertainty. This is because RST and DESAL140 combine to provide large amounts of water during periods of drier conditions and higher demand. The high energy consumption and total costs causes these options however, to fail after only 5 and 2 increments of uncertainty respectively.

Crossing of robustness curves provides valuable information. IPs 15 and 16 perform better than 17 and 18 in terms of storage susceptibility for the first 6 intervals of uncertainty. However, the steeper gradient of the robustness curves for IPs 17 and 18 shows that their storage susceptibility in this criterion is less affected by the increasingly harsh conditions at each successive interval of uncertainty. At the 7th interval of uncertainty IP 18 crosses the robustness curve for IP 15: beyond this interval of uncertainty IP 18's storage susceptibility is more robust to uncertainty than option 15's. This occurs again at the 8th interval of uncertainty where robustness curves for IP 15 and 17 cross. As conditions get harsher, IPs 17 and 18 perform comparatively better than IPs 15 and 16. Crossing robustness curves are also a feature of the service reliability performance criteria however, the crossings are oscillatory and do not denote a permanent shift.

IP 14 is only one increment of uncertainty less robust when compared to IPs 17 and 18. This IP includes UTR150, DESAL80 and SLARS. It is still an infrastructure heavy option, but it does not include the almost limitless capacity of the RST to maintain supplies. This slightly lower robustness comes with certain benefits: Fig. 9c and d show IP 14 is cheaper and less energy intensive than IPs 17 and 18. IPs 17 and 18 however outperform IP 14 in environmental performance, service reliability (L2 and L3) and storage susceptibility. The presence of SLARS in IP 18 makes it outperform IP 17 for storage susceptibility, environmental performance and service reliability (L2 and L3) at each increment of uncertainty at the expense of greater costs and energy consumption.

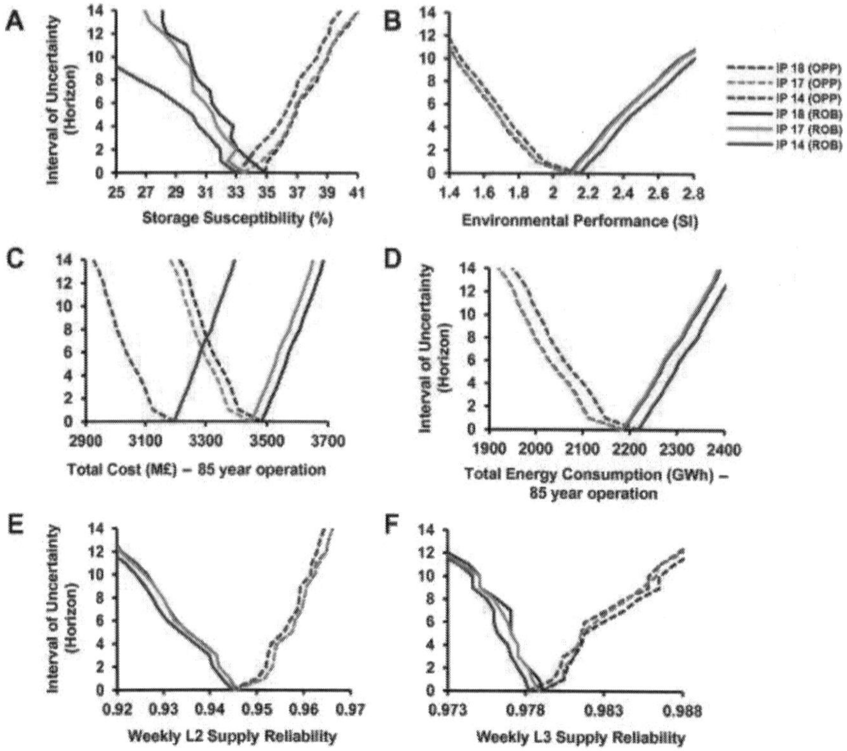

Figure 9. Opportuneness curves (dotted lines) for IPs 15, 16, 17 and 18 for storage susceptibility (a), environmental performance (b), cost performance (c) and energy consumption (d) and service reliability at L2 (e) and L3 (f). Curves show the improvement in performance for each option as we move incrementally towards increasingly more benign futures. The result show relative linearity between improvements in each performance measure and each successively more benign future. Robustness are also included to aid analysis of possible trade-offs (solid lines). Results show that performance rewards are comparable between the options except for total costs where IP 14 shows significant cost rewards (reductions) compared with IPs 17 and 18. These cost rewards are combined with lower rewards in the remaining performance metrics.

Results from the robustness analysis show IPs 14, 17 and 18 to be robust management options. Trade-offs between robustness in service reliability, environmental and storage susceptibility are at the expense of higher total costs and energy consumption, making it difficult to distinguish between these 3 IPs.

Opportuneness Function

The opportuneness function is derived from Eq. (6) and is the inverse of the robustness function. Opportuneness analysis quantifies the performance reward in each of our performance criteria Π_r (Res_o, $Cost_t$, $Energy_p$, Env_p, $Rel_{L2,L3}$) achieved in more benign futures derived from the right side of the Info-Gap uncertainty model. These futures become increasingly wetter, with lower demand and energy costs at each successive interval of uncertainty away from the best estimate:

$$\hat{\beta}(\Pi_r(Res_o, Cost_t, Energy_p, Env_p, Rel_{L2,L3}), q) = min\left\{h\left(\max_{u_i \in u(h)} \Pi_{Res_o, Cost_t, Energy_p, Env_p, Rel_{L2,L3}}(u_{1,2,3}, q)\right) \geqslant \Pi_r(Res_o, Cost_t, Energy_p, Env_p, Rel_{L2,L3})\right\} \quad (8)$$

Opportuneness analysis included 40 simulations each for IPs 14, 17 and 18. As with robustness analysis, curves are plotted for the performance of each option for each increment of uncertainty away from the central estimate. Opportuneness analysis explores system performance in futures that are less harsh. Whilst we are planning for system robustness, opportuneness can help choose between IPs with similar robustness (e.g. IPs 14, 17 and 18). Crossing curves denote when one IP gives greater reward than another. However, in opportuness analysis we are looking for the lowest curves with shallow gradients, indicating a high performance reward for small increments of uncertainty.

Opportuneness curves for IPs 14, 17 and 18 are plotted for each of our performance criteria (Fig. 9a–f). The corresponding robustness curves are also included to aid interpretation of possible trade-offs between robustness and opportuness. The curves are relatively linear for environmental performance, total cost and energy consumption. For service reliability at L2 and L3 there are step changes in calculated reward but the responses are comparable for all 3 IPs and do not result in crossing curves. Storage susceptibility opportuneness curves are quite linear, although Fig. 9a shows IP 17's opportuneness curve to have a shallower gradients than IPs 14 and 18 for the first two increments of uncertainty.

Without any crossing opportuneness curves the ranking of the IPs is unambiguous. IPs 17 and 18 consistently give similar levels of reward and IP 14 gives less reward for all criteria compared to IP 17 and IP 18 except total costs. IP 14 is significantly cheaper at each interval of uncertainty; on average it is approximately £3.3 M cheaper to operate each year than IPs 17 and 18 (based on it being £280 M cheaper over the 85 years of simulation). This greater cost reward must be traded-off against a lower overall robustness.

Based on this analysis IGDT promotes IP 18 (current infrastructure, RST and DESAL80) as the preferred water supply infrastructure portfolio. Compared to IP 14, IP 18 is more robust overall and performs better at each increment of uncertainty for all performance criteria except cost. IP 14's cost opportuneness is counter-balanced by its lower robustness. IPs 17 an 18 are similar; IP 17 is more robust to cost uncertainty but IP 18 outperforms IP 17 for all other criteria. IGDT leads us to select IP 18 as the preferred plan for a conservative planner. If less deviation from the future best estimate were to occur IP 14 would be more appealing for its robustness to cost increases in harsher futures and greater cost rewards in more benign futures.

DISCUSSION

RDM Discussion

An RDM analysis begins by the selection of a candidate strategy. In this case study, the RDM process began with a multi-criteria regret analysis. We generated a solution space of 311,080 simulations using a computationally efficient system simulator. The solution space included dire, mild and opportune plausible futures allowing us to compare IPs under a wide range of conditions. Each IP was compared to the best performing IP for a given set of input parameters. We used median relative regret to select IP 14 as the candidate strategy. IPs 13, 18 and 17 had the next lowest median regret.

The initial regret analysis also provided information on the relative performance of each IP. SLARS has minimal effect on the levels of regret and IPs with DESAL140 have higher regret in the cost and energy criteria than those with DESAL80. For storage susceptibility, service reliability and environmental performance however, DESAL140 improved regret in similar IPs.

The second step of the RDM framework involved a scenario discovery process using PRIM to isolate scenarios in which the candidate strategy is likely to fail. Two scenarios were identified: one which includes futures with dry summers and moderate-to-high demand and a second with dry summers, very dry winters and very low demand. In total 78% of failure points were covered in the three scenarios albeit with a high density of 95% (compared to the 55% failure density in the whole of the solution space). Given the spread of failure points around the solution space, PRIM was unable to produce scenarios with a better coverage to density ratio. This is a result of having multiple performance criteria that conflict; i.e. better environmental, storage and reliability performance often results in high energy consumption and high costs generating failures that are spread throughout the solution space.

A relative regret analysis of the simulations inside and outside vulnerability scenarios 1 and 2 showed that while IP 14 has the lowest median regret outside the vulnerability scenarios, IP 18 has the lowest median regret inside the vulnerability scenarios. The UTR150 in IP 14 and the RST in IP 18 provide the same amount of water to London when first activated during times of drought. However, because of how it is modelled the RST provides this release continuously and is unaffected by Thames area droughts (unlike the UTR).

From scenario 1 (which includes dry summers) and the regret analysis on scenarios 1 and 2, we identified IP 14 as vulnerable to failure because of its inadequate storage under droughts making it unable to compensate for the summer/winter hydrology disparity in climates that have dry summers. We identified that the RST is a useful, but an expensive and energy intensive alternative to address this disparity. We assumed the Severn basin will not experience drought at the same time as the Thames basin. A spatial drought correlation study would be required to establish the reliability of the RST and consider it in this study. This implies we cannot make the 'IP 14 or 18' decision for planners and that the relative strength of IPs that include the RST could be further lowered in regards to reliability. The price of energy was not found to affect the vulnerability of IP 14 because the majority of system failures come from the service reliability criteria.

In a full RDM analysis, decision makers could improve IP 14 and possibly IP 18 devising new IPs which could be resubmitted into the RDM process. The process would be repeated until a suitably robust IP is identified.

IGDT Discussion

The Info-Gap analysis was implemented using the fractional-error Info-Gap uncertainty model (Eq. (6)). This model characterised uncertain system input parameters and provided an interval-based approach to progressively sample harsher and more benign futures as compared to a best estimate.

The IGDT approach uses the vector *h* to incrementally sample the uncertain space. Because the three uncertain system parameters (inflows, demands, energy costs) are all scaled by *h* for each 'horizon' or increment of uncertainty, all parameters change proportionately; becoming increasingly harsher or more benign in robustness and opportuneness analysis respectively. Inflows lower whilst demands and energy costs increase at each successive interval of uncertainty. Although these parameters can be scaled and constrained within the Info-Gap uncertainty model to better represent natural relationships (Hall et al., 2011), conditions do change concurrently and proportionately. This has the advantage of requiring fewer simulations: only 40 simulations for each IP rather than a full grid search of 64,000 simulations for each IP (40 inflow, 40 demand and 40 energy intervals). One potential draw-back of the fractional-error IGDT approach is that it does not sample conditions where for instance; demand decreases but energy costs increase relative to our best estimate, or where inflow and demand are both lower than expected (vulnerable conditions as described by RDM's Scenario 2).

Results from our Info-Gap analysis show IP 18 is the most robust future infrastructure portfolio for maintaining adequate supply to London under harsher futures that are up to 5% dryer, with up to 2179 Ml/d potable water demand and energy costs up to 15.25 p/kW h. IP 18 maintains minimum system performance requirements for the greatest number of uncertainty intervals from our best estimate whilst maintaining water supplies to London (i.e. IP 18 fails as a result of environmental performance rather than service reliability). Plotting of robustness curves allows comparing the performance of different IPs for each metric at structured intervals of uncertainty around a best estimate. Robustness curves show IP 18 is the most robust to uncertainty in terms of storage susceptibility, service reliability and environmental performance. The steeper gradient of IP 18's robustness curves for storage susceptibility demonstrate there is less loss in performance at each successive interval of uncertainty when compared to IPs 15 and 16. The crossing of robustness curves at points 1 and 2 in Fig. 8b, marks the location where IP 18 becomes more robust than IP 15 and 16. The crossing of robustness curves results from the complex interactions between the supply infrastructure, environmental flow and demand reductions as dictated by the LTCD.

IGDT analysis suggests that IP 14 is also a good candidate strategy. This result is subject to the same caveats on RST as in RDM. Results from robustness and opportuneness analysis show IP 14 could be preferable to 18 if a decision maker could be more certain that the future will diverge less from the best estimate, or that the future will be more benign. Under both these conditions the improved cost performance of IP 14 compared to IP 18 for small deviations from the best estimate would potentially promote this as the most robust IP.

Thames Study Limitations

Several limitations affect the relevance of the results for real Thames region planning. The England and Wales water sector follows a 'twin-track' approach including both supply augmentation and demand management (e.g. water conservation, increased efficiency, metering, leakage reduction, communication campaigns, etc.). Our study only considers supply augmentation schemes to balance supply and demand; further research will bring in demand management options. The flow perturbation factors are applied to the historical flow time-series and do not take into account possible shifts in the hydrologic regime; they uniformly shift the intensity of flows (i.e. both low and high flows are equally augmented) and do not change the frequency or pattern of severe events (Prudhomme et al., 2010). Furthermore, the scaling of the flow factors in the IGDT analysis results in a compression or extension of the range of seasonal ranges of flows resulting in a further departure from the historical flow hydrology regime. Further research into methods that produce hydrologically consistent future flow time series is recommended and would strengthen the validity of the IGDT and RDM analyses. Uncertainty about the reliability of the River Severn Transfer should be considered, but without a water resource model of that area we could not consider it in this study. This would add a further dimension of uncertainty and significantly increase the computational burden of the planning problem. Further work could include this source of uncertainty and use sparser sampling methods for RDM. These improvements would increase the planning relevance of study results but would likely not significantly change methodological implications, the focus of this paper.

Complementarity of RDM and IGDT

Both RDM and IGDT helped assess the robustness of 20 different water supply infrastructure portfolios (IPs) to Knightian or 'deep' uncertainty in hydrological inflows, water demands and energy costs for a 2035 planning horizon. Due to differences in their approaches they reached similar but not entirely matching results. IGDT promotes IP 18 as the infrastructure portfolio of preference whereas RDM initially suggests IP 14. This observation leads to the main finding of this paper: because RDM and IGDT provide planners with different and complementary information they can be beneficially used together. We recommend joint use of IGDT and RDM for the planning of water resource systems as each method can help clarify the results of the other.

IGDT begins with a best estimate for each uncertain system input and then sequentially chooses values increasingly farther away from expected inputs. The best estimate of future conditions we used has a greater disparity between winter and summer flows, higher water demands and higher energy costs than the historical average. As a result the IGDT analysis, even at a zero interval of uncertainty, begins with the system under considerable stress before it considers less or more favourable conditions.

RDM samples from all combinations of the uncertain system parameters to identify conditions of system failure. RDM samples the 'extreme' futures as in IGDT but also more benign ones. In both IGDT and RDM, the robustness of simulated futures is assessed using multiple criteria such that IPs that are less infrastructure intensive (less costly and energy intensive) outperform more infrastructure intensive IPs in benign future states. These benign future states are taken into account during RDM's initial regret analysis and reduce the multi-criteria regret values of less infrastructure intensive options that perform reasonably well in storage reliability, storage susceptibility and environmental categories. IGDT does not consider these more benign futures and therefore less infrastructure intensive IPs do not perform well. For more benign futures, IP 14 has a lower median regret than IP 18, the more infrastructure intensive alternative while for more dire futures (Fig. 7), which more closely relate to the futures sampled in IGDT, IP 18 performs better. When only considering harsh futures, IGDT and RDM recommend the same future plan. Indeed, RDM's vulnerability scenario 1 (Table 6) shows IP 14 is vulnerable to failure under conditions that are only slightly harsher and even sometimes more benign (in the demand uncertainty dimension) than those that define the best estimate for the IGDT analysis.

Because IGDT assesses robustness of each option under progressively more dire futures, there is a tendency to favour IPs with infrastructure-intensive solutions. Using IGDT, it is difficult to jointly consider information about an IP's robustness to uncertainty and opportuneness for a performance reward. The regret analysis in our RDM implementation produced a single plot to characterise performance across all futures (Fig. 4). In this way RDM provides complementary information to an IGDT analysis on the suitability of future plans across a range of dire, benign and opportune futures. Our RDM application found solutions that are robust over a wide range of future states whilst our IGDT analysis identified solutions that are robust over dire futures and opportunistic over favourable futures. Taking into account the robustness of IPs for mild futures in water planning studies helps prevent decision makers from automatically favouring conservative (in this case infrastructure intensive) options. At the same time IGDT underlined the fact that our preferred option under RDM was quite vulnerable to dire futures.

IGDT robustness curves provide decision makers with useful visualisations that complement an RDM analysis. By plotting the performance of each option for each metric at each increment of uncertainty, trends in performance under increasingly dire or favourable conditions are revealed. This was seen for example in the service reliability performance criteria for IP 18 (Fig. 8e): initially IP 16 performs better but as conditions get worse IP 18 improves revealing it performs increasingly better under harsher conditions relative to IP 16. The IGDT plots show choosing IP 18 over IP 16 results in a cheaper and less energy intensive (Fig. 8c and d) system in harsher futures. IGDT robustness analysis also shows which performance criterion each IP fails at first whilst RDM does not track which criterion caused the failure. In this way we might consider environmental performance as a 'soft' performance rule which in reality could be reduced temporarily during droughts. IGDT tells us that under very harsh conditions, IP 18 fails environmental performance first, but if this is a shifting goal-post during droughts, IP 18 would still be acceptable (i.e. service reliability is not the first requirement to fail under harsh conditions).

The sampling in RDM meant a wide range of dire, benign and favourable conditions were considered during plan evaluation. RDM's scenario discovery approach identified the vulnerable sets of uncertain conditions for the preferred plan. These vulnerability scenarios were then used to check if other infrastructure portfolios could actually perform better under the rough conditions. Our RDM application showed IP 18 was the most robust option for more severe conditions; this was the preferred option from our IGDT analysis.

Our RDM implementation required 311,080 simulations and IGDT 1600 simulations. The supplementary runs required for IGDT suggest it is worthwhile to consider joint application. For example IGDT could be used along-side regret analysis to identify the initial candidate strategy, or to propose an alternative one for analysis in RDM's scenario discovery phase. Also, IGDT can be used to thin out the option sets considered in RDM; this could be useful when there are many possible initial options or plans. As discussed in Section 2.2. Info-Gap assumes analysts can provide a best estimate around which the uncertain parameter is assumed to lie and therefore does not explore the full uncertainty space. We found that complementing an IGDT approach with another such as RDM which samples future conditions more widely is an effective guard against potentially missing critical design conditions. Info-Gap can effectively be used to explore local robustness around selected scenarios in the uncertainty space. The overarching benefit of using RDM and IGDT methods is that they provide a structured approach to simulating systems under Knightian uncertainty. The methods select relevant system conditions to simulate and reveal how different plans perform under these conditions. Joint use widens the insights gained and helps verify and deepen the understanding that arises from using either method.

CONCLUSION

In this paper we formulated and applied the IGDT and RDM planning frameworks to a water resource management problem: expanding London's water supply system in the Thames basin, UK. We show that although the methods initially provide different capacity expansion recommendations, they produce broadly similar results but provide complementary information about the performance of different proposed infrastructure portfolios. Joint use of IGDT and RDM for the planning of water resource systems helps better understand the results of each method.

IGDT efficiently evaluates system performance under different options considering the most dire and opportune future conditions. The framework identifies which criteria cause failure of a future plan and allows to plot the performance of each option as future system inputs progressively deviate from an initial estimate. It may however ignore parts of a Knightian uncertainty space if the initial estimate ends up not being a good guess. RDM samples a wider set of combinations of the uncertain variables allowing a more holistic, albeit more computational intensive assessment of the future performance of different plans. Through RDM's scenario discovery combinations of uncertain inputs that lead to system vulnerabilities are identified. Regret analysis efficiently quantifies how different future plans fare over the vulnerable scenarios considering multiple performance criteria. RDM provides a structured framework for iterative refinement of future plans.

RDM and IGDT offer structured and insightful frameworks to plan complex systems with multiple requirements subject to multiple unknown future stresses. Joint use of the methods can make them more computationally efficient and maximise understanding by revealing how each method can skew results towards particular future plans.

ACKNOWLEDGEMENTS

The authors thank Ke Zhou for his help with early RDM work. The UK EPSRC funded the work through the ARCC-Water project (EP/G028540/1); A. Woods was partially supported by AECOM. Steven Wade and Brigitte von Christierson from HR Wallingford provided the climate change perturbation ratios. The authors thank Rob Wilby, the A.E. and a reviewer for comments that strengthened the article; the usual disclaimer applies.

REFERENCES

1. BBC, 2007. Water Firm may Lift Hosepipe Ban. (accessed 15.07.11). Ben-Haim, Y., 2001. Information-Gap Decision Theory. Academic Press, San Diego, CA.
2. Ben-Haim, Y., 2010. Info-Gap Economics: An Operational Introduction. Palgrave Macmillan, Basingstoke.
3. Beven, K., 2006. A manifesto for the equifinality thesis. J. Hydrol. 320 (1–2), 18–36.
4. Beven, K., Binley, A., 1992. The future of distributed models – model calibration and uncertainty prediction. Hydrol. Process. 6 (3), 279–298.
5. Bryant, B.P., Lempert, R.J., 2010. Thinking inside the box: a participatory, computerassisted approach to scenario discovery. Technol. Forecast. Soc. Change 77 (1), 34–49.

6. Christierson, B.V., Vidal, J.-P., Wade, S.D., 2012. Using UKCP09 probabilistic climate information for UK water resource planning. J. Hydrol. 424, 48–67.
7. Cole, G.A., Marsh, T.J., 2006. An historical analysis of drought in England and Wales. In: Demuth, S., Gustard, A., Planos, E., Scatena, F., Servat, E. (Eds.), Climate Variability and Change – Hydrological Impacts. Iahs Publication, pp. 483–489.
8. Cookson, A., Weston, A., 2008. Thames Water Modelling Project: Model 1a Final Report. Environment Agency.
9. Dessai, S., Hulme, M., 2007. Assessing the robustness of adaptation decisions to climate change uncertainties: a case study on water resources management in the East of England. Global Environ. Change – Human Policy Dimensions 17 (1), 59–72.
10. Eurostat, 2011. Electricity Prices for Industrial Consumers. (accessed 06.02.11).
11. Fredrich, A.J., 1975. Hydrologic Engineering Methods for Water Resources Development. Davis, California. Friedman, J.H., Fisher, N.I., 1999. Bump hunting in high-dimensional data. Stat. Comput. 9 (2), 123–143
12. Groves, D.G., Lempert, R.J., 2007. A new analytic method for finding policy-relevant scenarios. Global Environ. Change – Human Policy Dimensions 17 (1), 73–85.
13. Hall, J. et al., 2011. Robust climate policies under uncertainty: a comparison of robust decision-making and info-gap methods. Risk Anal..
14. Halrow Ltd., 2010. Thames Valley Stored Water System. (accessed 28.09.10). Hine, D., Hall, J.W., 2010.
15. Information gap analysis of flood model uncertainties and regional frequency analysis. Water Resour. Res., 46.
16. Hipel, K.W., Ben-Haim, Y., 1999. Decision making in an uncertain world: information-gap modeling in water resources management. IEEE Trans. Syst. Man Cybern. Part C – Appl. Rev. 29 (4), 506–517.
17. Johns, T.C. et al., 2003. Anthropogenic climate change for 1860 to 2100 simulated with the HadCM3 model under updated emissions scenarios. Clim. Dyn. 20 (6), 583–612.
18. Jones, A.E., 1983. Flow Gauging on the River Thames – The First 100 Years. National River Flow Archive, UK. Jones, M.A., Harris, S.J., Baxter, K.M., Anderson, M., 2005. The Streatham groundwater source: an analogue for the development of recharge enhanced groundwater resource management in the London basin. In: International Symposium on Management of Aquifer Recharge. United Nations Educational, Scientific and Cultural Organization, Berlin, Germany.

19. Knight, F.H., 1921. Risk, Uncertainty, and Profit. Houghton Mifflin, Boston, MA. Lempert, R., Groves, D., Popper, S., Bankes, S., 2006. A general, analytic method for generating robust strategies and narrative scenarios. Manage. Sci. 52 (4), 514– 528.
20. Lempert, R.J., Collins, M.T., 2007. Managing the risk of uncertain threshold responses: comparison of robust, optimum, and precautionary approaches. Risk Anal. 27 (4), 1009–1026.
21. Lempert, R.J., Groves, D.G., 2010. Identifying and evaluating robust adaptive policy responses to climate change for water management agencies in the American west. Technol. Forecast. Soc. Change 77 (6), 960–974.
22. Lempert, R.J., Popper, S.W.C.B.S., 2003. Shaping the Next One Hundred Years: New Methods for Quantitative, Long-Term Policy Analysis. RAND Corporation, Santa Monica, CA. Marsh, T., Cole, G., Wilby, R., 2007. Major droughts in England and Wales, 1800– 2006. Weather 62 (4).
23. Matrosov, E.S., Harou, J.J., Loucks, D.P., 2011. A computationally efficient open source water resource system simulator – application to London and the Thames Basin. Environ. Modell. Software 26 (12), 1599–1610.
24. McCarthy, M.A., Lindenmayer, D.B., 2007. Info-gap decision theory for assessing the management of catchments for timber production and urban water supply. Environ. Manage. 39 (4), 553–562.
25. Merrett, S., 2007. The Thames catchment: a river basin at the tipping point. Water Policy 9 (4), 393–404.
26. Moore, R.J., 1985. The probability-distributed principle and runoff production at point and basin scales. Hydrol. Sci. J. – J. Des Sci. Hydrol. 30 (2), 273–297.
27. Murphy, J.M. et al., 2007. A methodology for probabilistic predictions of regional climate change from perturbed physics ensembles. Philos. Trans. R. Soc. A – Math. Phys. Eng. Sci. 365 (1857), 1993–2028.
28. Murphy, J.M. et al., 2009. UK Climate Projections Science Report: Climate Change Projections, Met Office. Exeter, UK.
29. Nakicenovic, N., Swart, R. (Eds.), 2000. Special Report on Emissions Scenarios. A Special Report of Working Group III of the Intergovernmental Panel on Climate Change. Cambridge University Press, Cambridge, UK. Office for National Statistics, 2008. London: Population and Migration. (accessed 10.06.11).
30. OFWAT, 2010. A Study on Potential Benefits of Upstream Markets in the Water Sector in England and Wales, Birmingham, England. Prudhomme, C., Wilby, R.L., Crooks, S., Kay, A.L., Reynard, N.S., 2010.
31. Scenarioneutral approach to climate change impact studies: application to flood risk. J. Hydrol. 390 (3–4), 198–209.

32. Raluy, R.G., Serra, L., Uche, J., 2005a. Life cycle assessment of water production technologies – Part 1: life cycle assessment of different commercial desalination technologies (MSF, MED, RO). Int. J. Life Cycle Assess. 10 (4), 285–293.
33. Raluy, R.G., Serra, L., Uche, J., Valero, A., 2005b. Life cycle assessment of water production technologies – Part 2: reverse osmosis desalination versus the Ebro river water transfer. Int. J. Life Cycle Assess. 10 (5), 346–354. Savage, L.J., 1954. The Foundation of Statistics. Dover Publications.
34. Simon, H., 1959. Theories of decision-making in economic and behavioral science. Am. Econ. Rev. 49 (3), 253–283. Stokes, J., Horvath, A., 2006. Life cycle energy assessment of alternative water supply systems. Int. J. Life Cycle Assess. 11 (5), 335–343.
35. Thames Water, 2010. Revised Water Resources Management Plan, Thames Water, UK. WaterBriefing, 2006. Hopepipe Ban Remains while Drought Order Goes. (accessed 01.06.11).
36. Wilby, R., Greenfield, B., Glenny, C., 1994. A coupled synoptic hydrological model for climate-change impact assessment. J. Hydrol. 153 (1–4), 265–290.

CITATION

Evgenii S. Matrosov, Ashley M. Woods, Julien J. Harou, Robust Decision Making and Info-Gap Decision Theory for water resource system planning, Journal of Hydrology, Volume 494, 28 June 2013, Pages 43-58, ISSN 0022-1694, http://dx.doi.org/10.1016/j.jhydrol.2013.03.006.

CHAPTER 2

Impact Modelling of Water Resources Development and Climate Scenarios on Zambezi River Discharge

Harald Kling, Philipp Stanzel, Martin Preishuber

Pöyry, Hydropower and Renewable Energy, Water Resources Division, Laaer-Berg-Str. 43, 1100 Vienna, Austria

ABSTRACT

Study Region
The Zambezi River basin (1.4×10^6 km^2) in southern Africa, which is shared by eight countries and includes two of the World's largest reservoirs.

STUDY FOCUS

Impacts on future water resources in the Zambezi basin are studied, based on World Bank projections that include large scale irrigation and new hydropower plants. Also the impacts of climate change scenarios are analysed. Modelling challenges are the large basin area, data scarcity and complex hydrology. We use recent GPCC rainfall data to force a rainfall-runoff model linked to a reservoir model for the Zambezi basin. The simulations are evaluated with 60 years of observed discharge and reservoir water level data and applied to assess the impacts on historical and future discharges.

NEW HYDROLOGICAL INSIGHTS FOR THE REGION

Comparisons between historical and future scenarios show that the biggest changes have already occurred. Construction of Kariba and CahoraBassa dams in the mid 1900s altered the seasonality and flow duration curves. Future irrigation development will cause decreases of a similar magnitude to those caused by current reservoir evaporation losses. The discharge is highly sensitive to small precipitation changes and the two climate models used give different signs for future precipitation change, suggestive of large uncertainty. The river basin model and database are available as anopen-online Decision Support System to facilitate impact assessments of additional climate or development scenarios.

INTRODUCTION

"Stationarity is dead" – with this provocative statement Milly et al. (2008) raised a serious discussion for water resources planning in a changing world (see also the criticism byKoutsoyiannis, 2011, Lins and Cohn, 2011 and Matalas, 2012). Until recently, a common approach of hydrological engineers for water resources planning was to base the analysis on historic observations, while implicitly assuming that the past conditions are also representative of what to expect in the future. This approach is now more and more critically questioned due to non-stationarity observed in many hydrological variables and the possible impacts of climate change.

In addition to climate change, also development of water resources projects – such as dams for hydro-electric generation or irrigation projects – can have considerable impacts on discharge conditions, as summarized by mean flows, seasonality in flows or flow duration curve. In contrast to climate change – which is expected to be a transition on the time-scale of decades to centuries – development of individual water resources projects can mark an abrupt change for the hydrology in a basin.

In arid regions in Africa – where water is a limited resource – the impacts of climate change and water resources development are of particular concern, especially in international river basins. One example is the Zambezi basin that is shared by eight countries in the southern part of the African continent. Recent institutional strengthening with the establishment of the Zambezi Watercourse Commission (ZAMCOM, which came into force in 2011) aims at efficient and sustainable water resources management in the basin. In contrast to the Nile basin – where water resources are heavily exploited – irrigation projects in the basin are currently of limited importance, but large extensions are planned for the future. Two of the world's largest hydropower reservoirs (Kariba, Cahora Bassa) were already built in the middle of the 20th century at the Zambezi River, providing electricity for the region, but with significant downstream effects on river ecology.

The historic impacts of Kariba and Cahora Bassa dams on Zambezi discharge were analysed by Beilfuss and dos Santos (2001) and Matos et al. (2010) and there have been several studies proposing optimized operation rules to balance energy generation and ecological downstream impacts (e.g. Gandolfi and Salewicz, 1991, Tilmant et al., 2010,Beilfuss, 2010 and Mertens et al., 2013). There is concern that future development of large-scale irrigation projects may significantly reduce Zambezi River discharge, with negative impacts on hydropower and ecology (Hoekstra, 2003 and World Bank, 2010). On top of this, Zambezi discharge is also susceptible to possible future changes in climate (for a general overview see Beilfuss, 2012).

There are a few modelling studies that analysed future runoff conditions in the Zambezi basin under scenarios of climate change and water demand. This approach requires a fully-fledged hydrological modelling of the water fluxes in the basin and is therefore a considerable task, especially due to the fact that the models are set-up in a large, data-sparse region with a unique hydrology. Harrison and Whittington (2002) studied future energy generation at the proposed Batoka Gorge hydro-power plant at the Zambezi River below Victoria Falls. They modelled significant reductions in future discharge, albeit cautioning that "there is concern regarding the ability of the hydrological model to reproduce the historic flow". Yamba et al. (2011) applied the Pitman water balance model with selected climate scenarios to the full Zambezi basin to assess future energy generation at large hydro-power plants, obtaining results that show gradual reductions in discharge owing to climate change and increasing water demand. They show that their runoff simulations perform well in one tributary (Kabompo River), but do not present evaluations for the Zambezi River or the main tributaries. Beck and Bernauer (2011)modelled the combined changes in water demand and climate in 13 sub-basins of the Zambezi basin and the impact on mean water availability. They conclude that future climate change is of less concern, whereas population and economic growth as well as expansion of irrigated areas are likely to have important transboundary impacts due to significant decrease in water availability. They calibrated their hydrological model on long-term mean monthly discharge data, but do not present an evaluation of their discharge simulations with observed data.

Thus, the existing studies suggest that a reduction in future discharge is likely, but it is not clear how well the applied hydrological models perform for the simulation of Zambezi discharge, which raises questions about the modelling of discharge conditions under future climate change scenarios. Further, results of previous studies are difficult to compare due to different assumptions, models, time-periods and locations of interest. Therefore, the World Bank concluded in a recent study in the Zambezi basin that "additional detailed analysis is needed for assessing the impact of climate change" (World Bank, 2010, vol. 2, p. 83).

The objective of this study was to establish a well-calibrated hydrological model for the Zambezi basin, such that the model can be used with confidence for an assessment of the impacts of water resources development and climate change on Zambezi discharge. An important aspect of our study was a thorough evaluation of the historic simulations, to ensure that the model is capable of realistically representing the main input–output relationships of the system.

For future water resources development in the Zambezi basin we used scenarios of a highly detailed, recently published study (World Bank, 2010). On the other hand, there is a lack of detailed climate modelling for the African continent, where only data of coarse resolution general circulation models – with limited accuracy on the sub-basin scale – were readily available. For illustrative purposes we based our study on downscaled data of two well-known climate models, with contrasting projections about future precipitation.

The paper is structured as follows: After an introduction to the study area the data basis is presented. In the methods section we describe the river basin model, the calibration method and the scenario definitions. The results section includes an evaluation of simulation under historic conditions as well as results for simulation of future scenarios. This is followed by a discussion of results and possible sources of uncertainties. The paper ends with an outlook and conclusions.

STUDY AREA

This study focuses on the Zambezi basin (Fig. 1), which is the fourth largest river basin in Africa (after Congo, Nile and Niger) and covers 1.4 Mio km^2. As in other studies (e.g.Winsemius et al., 2006, Yamba et al., 2011 and Beck and Bernauer, 2011) we do not consider the Okavango River as a tributary of the Zambezi, even though in extremely wet years the Okavango system also partly discharges to the Zambezi. The basin is shared by eight countries: Zambia (41.9% of total area), Angola (18.2%), Namibia (1.1%), Botswana (1.5%), Zimbabwe (15.9%), Tanzania (2.2%), Malawi (7.5%), and Mozambique (11.6%). Typical vegetation types are woodland, grassland, and some agricultural areas, and elevation ranges from sea level to approximately 2500 m above sea level.

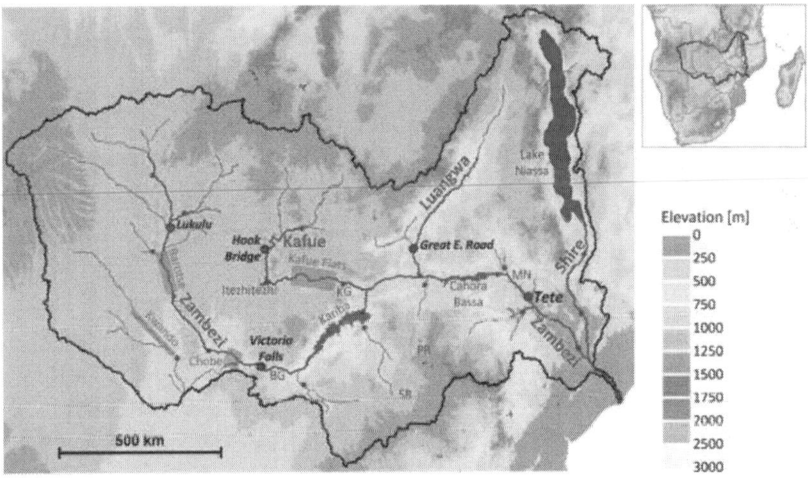

Figure 1. Zambezi River basin. Basin divides (black) and river-network (blue) derived from HydroSheds data set (Lehner et al., 2008). Names of main rivers in bold. Red circles: outlets of 27 sub-basins, with names of key locations in italic. Shaded boxes: natural floodplains and wetlands. Abbreviations for reservoirs: KG (Kafue Gorge), SR (Sanyati reservoirs), PR (Panhane reservoirs), BG (Batoka Gorge, planned), MN (Mphanda Nkuwa, planned). (For interpretation of the references to colour in this figure legend, the reader is referred to the web version of this article.)

The source of the Zambezi River is located at Kalene Hills in Zambia and travels roughly 2600 km to the south and east before discharging into the Indian Ocean at the Mozambican coast. Important tributaries from the north are the Kafue River, Luangwa River and Shire River, but there are no significant tributaries from the south. Floodplains and swamps (Barotse Floodplain, Chobe Swamps, Kafue Flats, Kwando Floodplain) are large, seasonally inundated areas of several thousand km^2. Lake Niassa – or also known as Lake Malawi – is located in the north-eastern part of the basin and is one of the world's largest freshwater lakes (570 km long, 30,000 km^2 surface area). There are also two large artificial reservoirs for hydropower generation at the Zambezi River (Lake Kariba with 5500 km^2 surface area and Lake Cahora Bassa with 2700 km^2). Lake Kariba is actually the world's largest artificial reservoir according to storage capacity (200,000 hm^3, GRanD global data set, Lehner et al., 2011).

Mean annual precipitation (MAP) is approximately 1000 mm/a, of which about 8% generates discharge and the remaining 92% is lost via evapotranspiration. The northern parts are wetter (MAP > 1250 mm/a) than the southern parts (MAP < 750 mm/a). During the dry season there is practically no precipitation. The wet season is during the austral summer and lasts from November to March. In most parts MAP is smaller than annual potential evapotranspiration, with a basin-wide average of 1600 mm/a. Mean discharge at the outlet of the basin is estimated to be approximately 3600 m^3/s, but discharge shows large seasonal and intra-annual variations. Seasonality in discharge is strongly controlled by seasonality in precipitation, but in addition also retention in large floodplains and swamps as well as artificial reservoirs affect the seasonal discharge. Zambezi floods travel several months from the headwaters in Zambia and Angola until reaching the lower reaches in Mozambique. In contrast, floods from the Luangwa tributary reach the Zambezi River within a few days, with similar peak flow as the upper Zambezi floods, but overall smaller flood volumes.

Even though in this study the whole Zambezi basin was modelled, in the paper we only report on the results for the Zambezi basin upstream of Tete (covering 1,103,400 km^2). Thereby, the Shire basin – with its specific hydrology due to the large impact of Lake Niassa – is excluded from the analysis.

DATA BASIS

Working in a large basin shared by several (eight) countries complicates acquisition of data. Therefore, the aim was to use as much as possible public data sources that are freely available.

Historic monthly climate data from 1901 to 2009 as spatial fields with a half degree (approximately 50 km) resolution were obtained from the following sources:

- Precipitation: Global Precipitation Climatology Centre (GPCC, version 5, published 2011), Deutscher Wetterdienst, Germany.
- Air temperature: Climate Research Unit (CRU, version TS 3.1, published 2011), University of East Anglia, UK.

The CRU temperature data in the Zambezi basin are based on interpolation from only few (approximately 10) stations, but in general interpolation of temperature data is assumed to be accurate due to strong correlation with elevation. Of more concern are the precipitation data, due to high spatial variability and the associated problems in interpolation from point measurements (see an assessment for the Zambezi region by Mukosa et al., 1995). In the Zambezi basin upstream Tete, GPCC is based on interpolation from approximately 100 stations during 1961–1990, but considerably fewer stations in other periods, especially after 1990 (Fig. 2). For such a large study area with more than 1 Mio km^2 this is a small number of stations given the high spatial heterogeneity of precipitation. However, the GPCC data set represents the best long-term observational data set available for the region. Note that the precipitation data of CRU – as used by, e.g. Beck and Bernauer (2011) – are based on only approximately half the number of stations as GPCC.

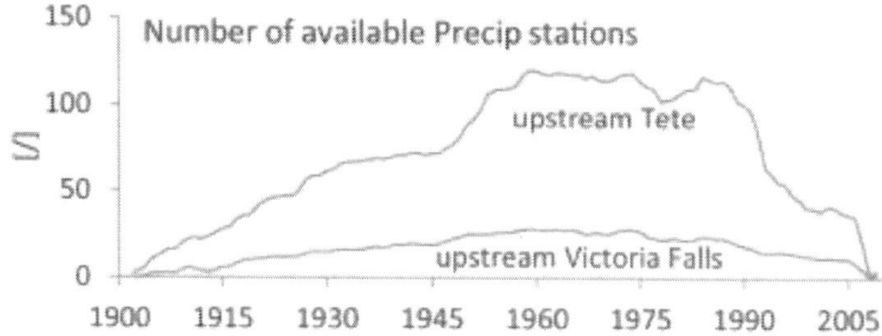

Figure 2. Availability of precipitation stations. GPCC data set from 1901 to 2009. Upper line: number of stations in the Zambezi basin upstream of Tete. Lower line: number of stations in the Zambezi basin upstream of Victoria Falls.

Long-term mean monthly potential evapotranspiration (mPET) data were obtained from the CLIMWAT data set of FAO for 30 stations in the region. The Penman–Monteith method (Monteith, 1965) was used in the CROPWAT model of FAO to calculate the sensitivity of mPET to changes in temperature. It was found that for an increase in temperature by +1 °C there is an increase in mPET by +2.5%, with insignificant differences in this factor between stations and months. Thus, this relationship is also used for preparing potential evapotranspiration time-series from historic and future (projected) temperature data (see equation in Appendix).

Climate scenario data about future precipitation and temperature were obtained from the recently finished EU WATCH project (WATer and global CHange, published 2011, http://www.eu-watch.org). In the WATCH project, daily data of GCMs (General Circulation Models, or Global Climate Models) were downscaled with quantile mapping with observed data of 1960–2000 (Piani et al., 2010) to a half degree spatial resolution. We applied an additional, small bias correction (linear scaling, see e.g. Lenderink et al., 2007) to aggregated monthly data, such that the GCM data matched the climatology 1961–1990 of the GPCC precipitation data and CRU temperature data. In this paper we report on the results with two climate models for the IPCC A2 emission scenario (high emissions), as summarized in Table 1.

Table 1. Climate model data of the WATCH project for the Zambezi basin upstream of Tete. A2 emission scenario.*P*: mean annual precipitation. *T*: mean annual air temperature. Δ*P*: change in precipitation relative to 1961–1990. Δ*T*: change in air temperature relative to 1961–1990.

Institute	GCM	Period	*P* [mm/a]	*T* [°C]	Δ*P* [%]	Δ*T* [°C]
CNRM[a]	CNRM-CM3	1961–1990	912	21.8	–	–
		2021–2050	967	23.5	6	1.7
		2071–2100	980	26.6	7	4.8
MPI[b]	ECHAM5/MPIOM	1961–1990	912	21.8	–	–
		2021–2050	892	23.2	−2	1.4
		2071–2100	862	26.5	−5	4.7

[a]CNRM: Centre National de Recherches Météorologiques (France).
[b]MPI: Max-Planck-Institut für Meteorologie (Germany).

Observed time-series of monthly discharge was obtained for 22 gauges. As Hughes et al. (2010) point out, the discharge measurements and climate data in southern Africa are "subject to unknown errors of varying degrees". Especially for discharge data plausibility checks (double-mass curves, upstream versus downstream comparisons) yielded ambiguous results. The reliability of discharge data appeared to change significantly over time, with each gauge having its own peculiarities. Therefore, in this paper we only report results for five gauges at key locations:

- Zambezi River at Lukulu (catchment area of 212,600 km^2): Zambezi headwaters, measurements available since 1954.
- Zambezi River at Victoria Falls (519,400 km^2): Upper Zambezi, discharge data of this gauge are assumed to be reliable and are available since 1908.
- Kafue River at Kafue Hook Bridge (96,200 km^2): Upper Kafue River, measurements available since 1908. Data of this gauge were obtained at the end of the study for an independent evaluation (not used for calibration).
- Luangwa River at Great East Road (142,000 km^2): Almost full Luangwa basin, measurements available since 1949.
- Zambezi River at Tete (1,103,400 km^2): Lower Zambezi River, measurements available since 1952, but there are implausible differences with nearby gauges. Further, data obtained from different providers for the same location and period frequently differed by up to 100 % (or 600 m^3/s) during low flow conditions.

Fig. 3 gives a summary of the acquired data by showing long-term trends for precipitation, air temperature and discharge. Historic precipitation data before 1930 and after 1990 should be interpreted with caution due to low availability of stations (see Fig. 2). The historic precipitation data show large inter-annual variability, but no clear trend. Climate model data show small trends, but with different signs according to the analysed model. In contrast, the temperature data show a clear warming trend after 1980, which corresponds with the changes on the global scale (IPCC, 2007). The climate model data project that warming continues throughout the 21st century. Annual discharge data of the Upper Zambezi at Victoria Falls exhibit large inter-annual variability – ranging between 400 m^3/s in dry years to 2300 m^3/s in wet years. There is a cyclic behaviour of Zambezi discharge, with above average flows during 1950–1980 (Mazvimavi and Wolski, 2006), which corresponds to small long-term variations in the precipitation data (for a discussion of multi-decadal climate variability in southern Africa see Tyson et al., 2002).

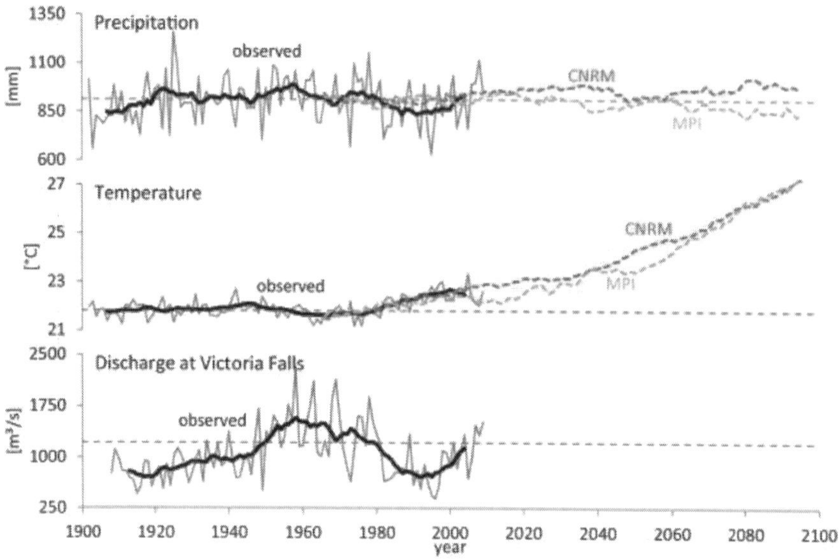

Figure 3. Long-term climate trends. Annual precipitation amount, mean temperature and mean discharge in the 20th century (observations) and 21st century (climate model projections). Precipitation and temperature data for Zambezi basin upstream Tete. Discharge data for upper Zambezi River at Victoria Falls. Thin black lines: observations in individual years. Bold black lines: 11-year moving average of observations. Dashed horizontal lines: mean of observation 1961–1990. Dashed coloured lines: 11-year moving averages of projections by climate models CNRM and MPI. (For interpretation of the references to colour in this figure legend, the reader is referred to the web version of this article.)

METHODS

In this study a river basin model – consisting of a water balance model and a water allocation model – was calibrated with historic data. The river basin model was then applied for selected scenarios to analyse the impact of water resources development and climate change on Zambezi River discharge. The following sections describe the water balance model, the water allocation model, the calibration method and the scenario definitions.

Water Balance Model

The water balance model simulates the precipitation-runoff process in 27 sub-basins of the Zambezi basin. The size of the sub-basins ranges between 10,300 and 132,300 km^2, with a mean size of 50,900 km^2. The sub-basin outlets are depicted in Fig. 1.

In each sub-basin the same model concept is applied (Fig. 4, left). This model was already used in several climate change impact studies in central Europe (e.g. Stanzel and Nachtnebel, 2010 and Kling et al., 2012). Similar model structures proved to be successful for the Zambezi (e.g. Winsemius et al., 2008). Inputs are monthly precipitation and potential evapotranspiration. Precipitation can be stored and evaporated from the interception storage. The remaining water falling on the ground is either stored in the soil or generates runoff as an exponential function of soil moisture (HBV-type concept,Bergström, 1995). Evapotranspiration from the soil depends on soil moisture and potential evapotranspiration. Generated runoff is split into a fast component (surface flow) and a slow component representing base flow (simulated as a linear reservoir). In general monthly time-steps are used, but the interception and soil modules internally use descretizations into daily time-steps to account for intra-monthly variability (interception/evaporation of individual rainfall events; inter-dependence of soil moisture, evapotranspiration and runoff generation). The model equations are listed in the Appendix.

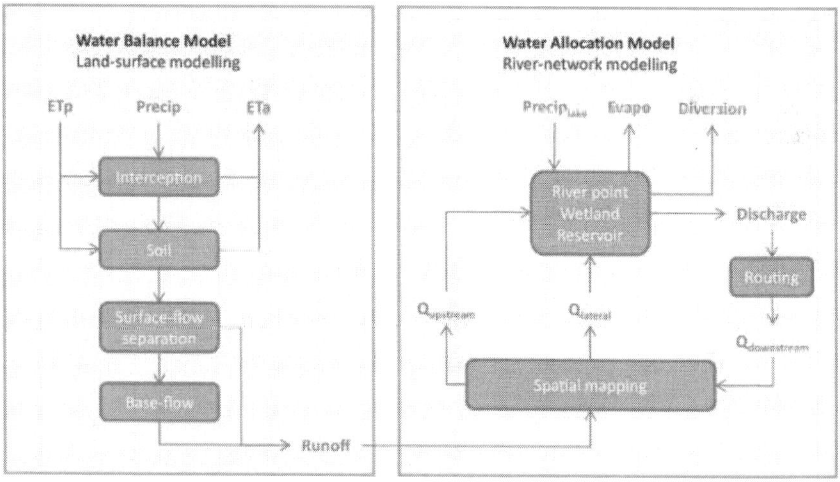

Figure 4. Conceptual model structure. River basin model consisting of water balance model (left) and water allocation model (right). Precip: precipitation in mm. ETp: potential evapotranspiration in mm. ETa: actual evapotranspiration in mm. Runoff: runoff-depth in mm. Q_{upstream}: upstream inflow in m^3/s. Q_{lateral}: lateral inflow in m^3/s. Precip$_{\text{lake}}$: precipitation on open water body in m^3/s (is zero for River points). Evapo: evaporation from open water body in m^3/s (is zero for River points). Diversion: withdrawal of water in m^3/s. Discharge: river discharge in m^3/s. $Q_{\text{downstream}}$: routed downstream discharge in m^3/s.

Water Allocation Model

The water allocation model aggregates runoff of the water balance model along the river-network to compute discharge and was developed new for this study. Even though the inputs and outputs have a monthly temporal resolution, daily time-steps are used for the internal computations.
The model considers the following elements (Fig. 4, right):

- River points: Used for querying discharge at locations of interest.
- Uncontrolled reservoirs: Wetlands, floodplains, and lakes.
- Controlled reservoirs: Large, artificial reservoirs impounded by dams.
- Diversions: Consumptive use (e.g. withdrawals for irrigation).
- Routing: Simple lag method for discharge routing.

The standard set-up of the water allocation model consists of 38 computation points (see also Fig. 1):

- 27 river points at the sub-basin outlets.
- 5 uncontrolled reservoirs: Barotse Floodplain (3500 km^2), Kwando Floodplain (3000 km^2), Chobe Swamps (2000 km^2), Kafue Flats (1900 km^2) and Lake Niassa (30,000 km^2).
- Controlled reservoirs, as listed in Table 2.

Table 2. Reservoirs considered in this study. The last two reservoirs are currently in the planning stage.

Reservoir	Country	Capacity [hm^3]	Water surface [km^2]	Start year
Kariba	Zambia/Zimbabwe	200,000	5500	1959
Cahora Bassa	Mozambique	85,000	2700	1975
Itezhitezhi	Zambia	7000	375	1978
Kafue Gorge	Zambia	950	13	1971
Sanyati reservoirs	Zimbabwe	450	46	1957
Panhane reservoirs	Zimbabwe	1200	125	1976
Batoka Gorge	Zambia/Zimbabwe	1700	26	2024
Mphanda Nkuwa	Mozambique	2300	100	2024

Additional computation points were inserted to query discharge at locations of interest (e.g. Kafue Hook Bridge) and to study the impact of planned reservoirs (Batoka Gorge, Mphanda Nkuwa).

A key characteristic of controlled and uncontrolled reservoirs is the relationship between storage (hm^3), water surface (km^2), water level (m) and release (m^3/s). At uncontrolled reservoirs the release is a direct function of storage. At controlled reservoirs the release depends on a prioritization of water:

- Environmental flow as a function of month.
- Diversions (e.g. for irrigation) as a function of month.
- Desired release (e.g. for hydropower) as a function of water level (reservoir zoning concept).
- Guide curve operation (e.g. for flood control) as a function of month.

The water surface area may show large seasonal fluctuations especially at natural floodplains, thereby affecting evaporation fluxes. Evaporation is computed as the potential evapotranspiration increased by 5% (according to FAO 56, Allen et al., 1998) and multiplied by the water surface area. Other fluxes at reservoirs include upstream inflows, lateral inflows, and precipitation on the water body. Overall, the model is able to mimic the most important reservoir operation characteristics, as, e.g. also used by the well-known HEC-ResSim model.

Model Calibration

The calibration of the river basin model combined methods of a priori estimation (literature review), sensitivity analysis, automatic optimization and manual parameter adjustments with the overall objective to obtain simulations that are consistent with available observations – i.e. observed discharge data measured at gauges and observed water levels in large reservoirs.

The main focus was on calibration of parameters of the water balance model. Initial parameter estimates were based on previous studies that give valuable insights into the hydrological behaviour of the Zambezi basin (Scipal et al., 2005, Winsemius et al., 2006,Winsemius et al., 2008 and Meier et al., 2011). Three parameters were calibrated, whereas the other parameters were manually set to the same values in all sub-basins (see Table 7 in Appendix). Sub-basins with no observed discharge data available for optimization were assigned parameter values of neighbouring sub-basins. The same applied to the downstream sections (e.g. Zambezi at Tete) with no reliable gauge data. The three optimized parameters that vary between (groups of) sub-basins include:

- Soil storage capacity.
- Soil exponent for runoff generation.
- Fraction for separation of surface flow.

The first two parameters affect storage of rainfall in the soil for evapotranspiration and thereby control mean volume of flow. Further, they control how long it takes (up to several months) in the rainy season before the soils are sufficiently wet to enable runoff generation (see also Scipal et al., 2005 and Meier et al., 2011). The third parameter defines the fractions of runoff representing surface flow – which leaves the sub-basin within the same month – and base flow with a delayed response controlling dry season discharge.

Observed discharge data of the period 1961–1990 at 14 gauges were used to automatically calibrate these three parameters of the water balance model with the Shuffled Complex Evolution search algorithm (Duan et al., 1992). As objective function we used a slightly modified version of the *KGE*-statistic (Gupta et al., 2009; modified according to Kling et al., 2012):

$$KGE' = 1 - \sqrt{(r-1)^2 + (\beta-1)^2 + (\gamma-1)^2} \qquad (1)$$

$$\beta = \frac{\mu_s}{\mu_o}$$

$$\gamma = \frac{CV_s}{CV_o} = \frac{\sigma_s/\mu_s}{\sigma_o/\mu_o}$$

where KGE' is the modified version of the *KGE*-statistic (dimensionless), r is the correlation coefficient between simulated and observed discharge (dimensionless), β is the bias ratio (dimensionless), γ is the variability ratio (dimensionless), μ is the mean discharge in m^3/s, CV is the coefficient of variation (dimensionless), σ is the standard deviation of discharge in m^3/s, and the indices s and o represent simulated and observed discharge values, respectively. KGE', r, β and γ have their optimum at unity. For a full discussion of the *KGE*-statistic and its advantages over the often used Nash–Sutcliffe Efficiency (NSE, Nash and Sutcliffe, 1970) or the related mean squared error see Gupta et al. (2009).

The *KGE*-statistic offers interesting diagnostic insights into the model performance because of the decomposition into correlation (r), bias term (β) and variability term (γ). In this paper we use this decomposition of the model performance to report on the evaluation of discharge simulations at five key locations within the Zambezi basin in the calibration period 1961–1990 as well as in the independent evaluation period 1931–1960. Because of the long observed discharge time-series these statistics were also computed at the gauge Kafue Hook Bridge, even though this gauge was not included in the original set-up of the model.

In addition to the parameters of the water balance model, there were also a large number of parameters that had to be specified for the water allocation model. These parameters were not calibrated in a classical sense. Most of them could be estimated a priori using data of reports (reservoir characteristics, reservoir operation rules effective during the 2000s), analysis of spatial data (elevation, vegetation) with Geographic Information System (wetland characteristics), and daily discharge data of upstream versus downstream gauges (lag-parameter for routing). Release rules of reservoirs were further refined by analysis of observed reservoir outflows during dry periods. Storage–discharge relationships of floodplains and wetlands in the upper Zambezi basin were determined manually after sensitivity tests.

Scenario Definition

We used the scenarios listed in Table 3 to separately assess the impact of water resources development and climate change on discharge in the Zambezi basin. Such a scenario approach required to first define a baseline scenario, for comparison against all the other scenarios. In the case of the Zambezi basin, the simulation of historic conditions – as in the calibration and evaluation periods – was not suitable for such a baseline scenario, due to abrupt changes in discharge conditions caused by the building of large dams and the subsequent filling of the reservoirs over several years, which temporarily reduced downstream discharge significantly. Therefore, a separate “Baseline” scenario was defined using observed climate data of the period 1961–1990 but including all existing large reservoirs as of year

2010 (Table 2). For this scenario the reservoirs are always under operation, regardless of commissioning date. Further, this scenario also includes existing irrigation withdrawals according to World Bank (2010), where for each sub-basin, a mean monthly irrigation demand was available. These irrigation withdrawals were not included in the "Calibration" and "Evaluation" scenarios because of lack of information about start years of individual irrigation withdrawals and generally low irrigation levels.

Table 3. Overview about scenarios considered in this study, listing mean annual values of Zambezi basin upstream of Tete for precipitation (*P*), air temperature (*T*), and irrigation demand (according to World Bank, 2010). Reservoir scenarios include "historic" (considers start year as listed in Table 2 "current" (all reservoirs with start year before 2010), and "future" (also planned reservoirs). Irrigation data also includes withdrawals of 15 m^3/s for water supply of the city Lusaka in Zambia.

Scenario name	Climate scenario				Development scenario	
	Source	Data period	P [mm /a]	T [°C]	Reservoirs	Irrigation [m^3/s]
Historic scenarios						
Calibration	Observed	1961–1990	912	21.8	Historic	15
Evaluation	Observed	1931–1960	933	21.9	Historic	15
Reference scenario						
Baseline	Observed	1961–1990	912	21.8	Current	85
Water resources development scenarios						
Pristine	Observed	1961–1990	912	21.8	None	0
Moderate development	Observed	1961–1990	912	21.8	Future	203
High development	Observed	1961–1990	912	21.8	Future	651
Climate model scenarios						
CNRM near future	CNRM	2021–2050	967	23.5	Current	85
MPI near future	MPI	2021–2050	892	23.2	Current	85
CNRM far future	CNRM	2071–2100	980	26.6	Current	85
MPI far future	MPI	2071–2100	862	26.5	Current	85
Climate sensitivity scenarios						
P −10%	(Observed)	1961–1990	821	21.8	Current	85
P +10%	(Observed)	1961–1990	1003	21.8	Current	85
T +2 °C	(Observed)	1961–1990	912	23.8	Current	85
T +4 °C	(Observed)	1961–1990	912	25.8	Current	85

Three different development scenarios were considered for water resources management. The "Pristine" development scenario includes neither reservoirs nor diversions, thus representing undisturbed conditions in the Zambezi basin. The "Moderate" and "High" development scenarios represent different levels of irrigation according to World Bank. For each scenario the corresponding mean monthly irrigation diversions are applied to the 38 computation points of the model. Moderate development includes identified irrigation projects that may be realized within the next decades, whereas High development includes all theoretically possible irrigation projects. For both scenarios the planned reservoirs Batoka Gorge and Mphanda Nkuwa were considered to be under operation. Several other, smaller planned reservoirs were not considered. For all three development scenarios the observed climate data of the period 1961–1990 were used.

Four climate scenarios were considered based on data of two climate models (CNRM, MPI) and two time periods. "Near future" was defined as 2021–2050 and "far future" was defined as 2071–2100. For these scenarios the same development scenario was specified as for the Baseline scenario.

In addition to the climate scenarios based on GCM data, further scenarios were defined for climate sensitivity analysis. Observed climate data of the period 1961–1990 were modified by increasing/decreasing precipitation by 10%, as well as increasing temperature by +2 °C and +4 °C. Again the same development as in the Baseline scenario was used.

For the sake of brevity and clarity we do not present scenarios that are combinations of different levels of development and climate projections. One obvious combination would be to assess the impact of Moderate development in conjunction with climate model projections for the near future. However, the current climate model projections are highly uncertain which we show in the results section. Therefore, little could be learned from additional scenario combinations.

RESULTS

First we report on the simulation results for discharge under historic conditions and the related performance of the river basin model. Subsequently, results of the scenario simulations for the pre-defined development and climate change scenarios are presented.

Simulation of Historic Conditions

This section gives insights into the historical hydrological conditions of the period 1961–1990 in the Zambezi basin, as observed and modelled.

Fig. 5 shows a comparison of simulated and observed monthly hydrographs for the Upper Zambezi River at Victoria Falls and the Zambezi River at Tete. With the exception of a few years, the simulated discharge closely matches the observed discharge at Victoria Falls. The differences are larger for the simulation of discharge at Tete, but still the general characteristics are simulated well. From Fig. 5 it is clear that the hydrograph at Victoria Falls represents undisturbed river flows with typical seasonality, whereas the hydrograph at Tete is impacted by the operation of the large Kariba and Cahora Bassa reservoirs. For example, during the 1980s there was no typical seasonality in discharge due to constant releases from Kariba reservoir in dry periods and flood attenuation in wet periods. From 1975 to 1977 the simulations deviate considerably from the observed discharge. During this period Cahora Bassa reservoir was first filled and the operation rules imposed on the model do not reflect the actual operations in this period well. During the 1980s water levels in Cahora Bassa reservoir were affected by the armed conflict in Mozambique. The reservoir was not run with normal operations from 1981 to 1998 because transmission lines from the hydropower plant were destroyed.

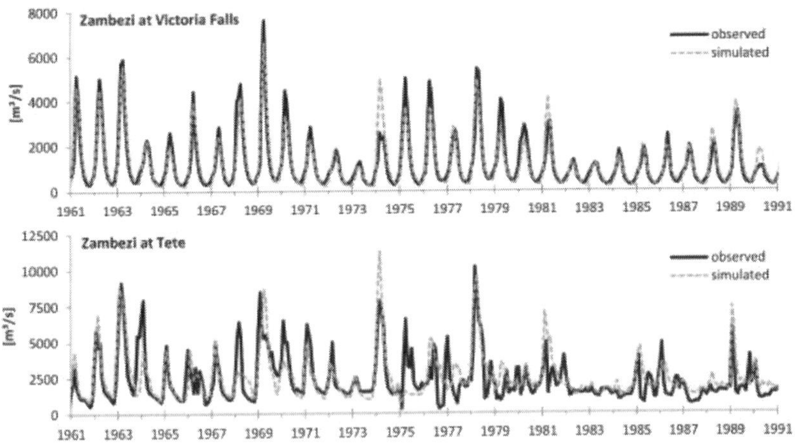

Figure 5. Evaluation with discharge data. Simulated and observed monthly hydrographs 1961–1990. Top: upper Zambezi River at Victoria Falls. Bottom: Zambezi River at Tete.

The simulation of the operation of Kariba reservoir – which is the largest reservoir in the basin and twice as large as Cahora Bassa – is evaluated next. Fig. 6 shows a comparison of simulated and observed water levels. Kariba dam was completed in 1959 and the filling of the reservoir lasted until 1963, which is simulated well (Fig. 6, left side). In the two-year period from the middle of 1961 to the middle of 1963 the volume of water stored in the reservoir increased by 100 km^3, resulting in an average reduction of downstream discharge of approximately 1600 m^3/s (= 100 km^3/2 years). After this first complete filling of the reservoir the water level was held at a lower level from 1964 to 1973 than in later periods. Release decisions were also affected by electricity generation, where the installed capacity of turbines increased over time. From 1974 onwards the simulated water levels closely match the observed water levels. From 1981 to 1984 the water level dropped because of low inflows but constant, higher releases. During this four-year period the volume of stored water decreased by 60 km^3, thereby increasing downstream discharge by an average of approximately 500 m^3/s. In the last two years of Fig. 6 (1989 and 1990) water levels are over-estimated because of too high simulated inflows (see discharge simulation at Victoria Falls in Fig. 5). Overall, the general impact of reservoir operation is simulated sufficiently well, even though there may be deviations in individual years.

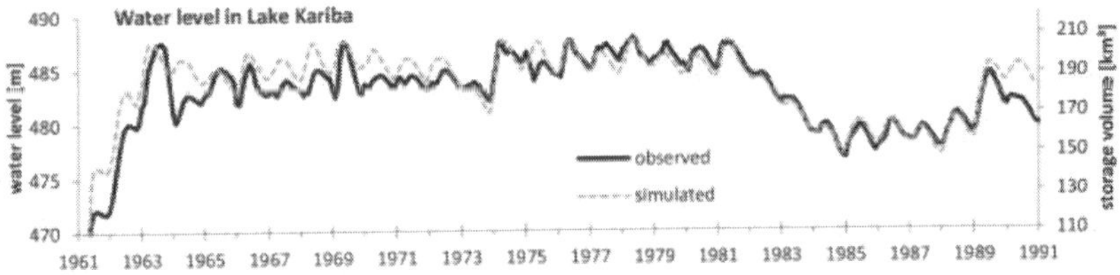

Figure 6. Evaluation with water level data. Simulated and observed monthly water levels in Kariba reservoir. Period 1961–1990.

In addition to the reservoir simulation discussed above, of key interest is also the simulation of undisturbed discharge conditions at the three main tributaries: Upper Zambezi River, Kafue River, and Luangwa River. Fig. 7 shows that both the seasonality in discharge and the overall distribution of discharge (monthly flow duration curve) are simulated well. Mean annual discharge of the Upper Zambezi is with 1200 m^3/s much larger than for the Kafue River (370 m^3/s) and Luangwa River (600 m^3/s). A separate evaluation in the ten wettest and ten driest years of 1961–1990 for the Upper Zambezi River shows that the model accurately simulates the different discharge conditions in wet and dry years (Fig. 8). Mean annual discharge in wet years is with 1700 m^3/s more than twice as large as in dry years (800 m^3/s), even though differences in annual precipitation are not as pronounced with values of 1060 mm/a in the 10 wettest years versus 820 mm/a in the 10 driest years. This means that the percentage change between wet and dry years is for discharge approximately four times larger than for precipitation, highlighting the high sensitivity of discharge to precipitation.

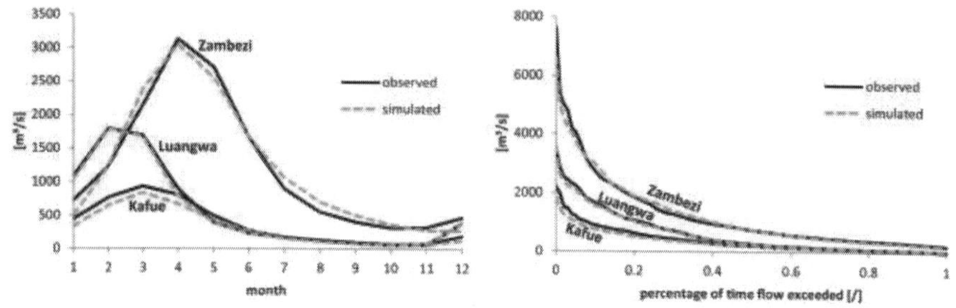

Figure 7. Simulated and observed discharge conditions of key tributaries. Zambezi: upper Zambezi River at Victoria Falls. Luangwa: Luangwa River at Great East Road. Kafue: Kafue River at Kafue Hook Bridge. Period 1961–1990. Left: seasonality in discharge. Right: monthly flow duration curve.

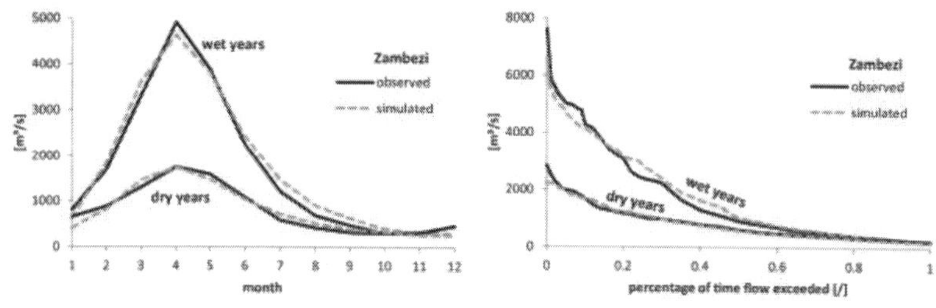

Figure 8. Simulated and observed discharge conditions in wet and dry years. Upper Zambezi River at Victoria Falls. Separate evaluation in 10 driest and 10 wettest years according to precipitation data of the period 1961–1990. Left: seasonality in discharge. Right: monthly flow duration curve.

To better understand the processes governing the generation of discharge Fig. 9 shows the simulated seasonal water balance averaged over the land-surface of the Zambezi basin upstream of Tete (water bodies of wetlands and reservoirs, as well as the effect of routing, are excluded from this analysis). Runoff-depth is only a small fraction in relation to the other components of precipitation, actual evapotranspiration and storage change (which gives the cumulative changes of water stored as soil-moisture and ground-water). In the first months of the rainy season (October–December) no runoff is generated because all of the precipitation is either stored in the ground – indicated by positive values of storage change – or lost via actual evapotranspiration. Actual evapotranspiration is however considerably smaller than potential evapotranspiration due to dry soils. This changes towards the end of the rainy season (February, March) when soils become wet and actual evapotranspiration is similar to potential evapotranspiration. During this period with wet soils runoff is eventually generated from precipitation, but the overall amounts of runoff are still an order of magnitude smaller than the other water balance components. After the end of the rainy season in April runoff is still significant due to base flow. Actual evapotranspiration becomes larger than precipitation – which is basically zero during the dry season from May to September – resulting in drying up of soils indicated by negative storage change. The peak in potential evapotranspiration in September and October – caused by hot, dry and windy conditions – has no direct impact on actual evapotranspiration due to lack of water.

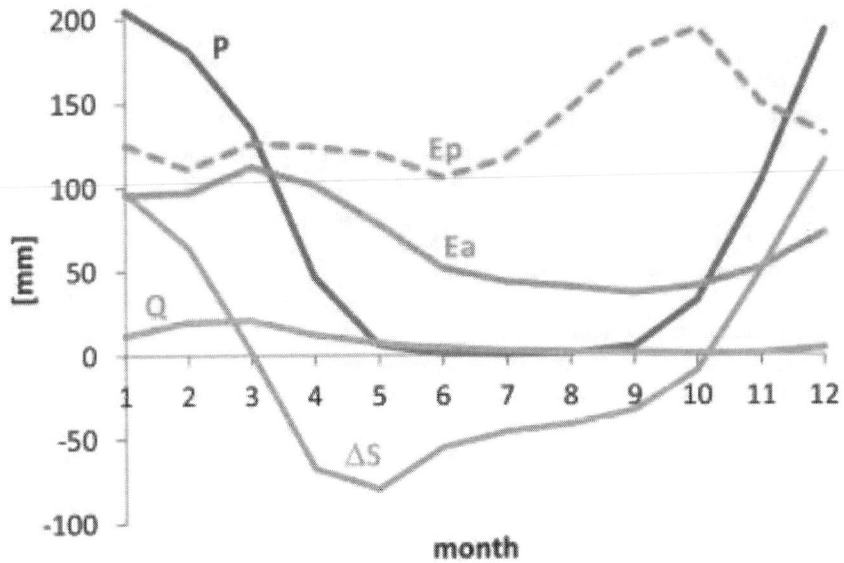

Figure 9. Simulated seasonal water balance of the Zambezi basin upstream of Tete. Variables represent mean monthly amounts of the period 1961–1990. P: precipitation. Ep: potential evapotranspiration. Ea: actual evapotranspiration. Q: runoff-depth (does not include the impact of routing, wetlands, reservoirs, diversions). ΔS: storage change. Water balance equation: $P = \text{Ea} + Q + \Delta S$.

Model Performance Statistics

In addition to the evaluation based on visual comparisons presented in the previous section, we also report on the model performance statistics for the calibration period (1961–1990) and the independent evaluation period (1931–1960).

Table 4 lists the performance statistics for discharge simulation at key locations. At some gauges data are available only in a limited number of years during the evaluation period, but time-series are mostly complete in the calibration period. In general the model performance is high in both periods, with a few exceptions as discussed further below. In most cases the correlation is above 0.90 and the Nash–Sutcliffe efficiency is above 0.80. This applies for the calibration period as well as the independent evaluation period.

Table 4. Model performance for monthly discharge simulation at key locations in the Zambezi basin. Years: number of years with available discharge observations. *KGE'*: modified *KGE*-statistic (see Eq. (1)). *r*: correlation coefficient. *β*: bias ratio. *γ*: variability ratio. NSE: Nash–Sutcliffe Efficiency. All performance statistics are dimensionless.

		Calibration period 1961–1990					Evaluation period 1931–1960					
River	Gauge	Years	*KGE'*	*r*	*γ*	NSE	Years	*KGE'*	*r*	*β*	*γ*	NSE
Zambezi	Lukulu	29	0.9	0.93	1.08	0.85	7	0.87	0.97	1.1	0.92	0.92
Zambezi	Victoria Falls	30	0.92	0.94	0.96	0.88	30	0.79	0.91	1.19	0.99	0.73
Kafue	Hook Bridge	30	0.83	0.94	0.92	0.85	29	0.81	0.92	1.01	0.83	0.84
Luangwa	Great E. Road	28	0.93	0.94	0.99	0.88	11	0.84	0.93	1.07	0.87	0.86
Zambezi	Tete[a]	30	0.71	0.74	0.9	0.48	9	0.7	0.95	1.28	1.1	0.63

[a]Zambezi discharge data at Tete are not reliable, and in addition are affected by changing operation rules

Even though performance statistics in Table 4 are also listed for the gauge Tete, it has to be considered that the reported observed discharge data for this gauge are of limited accuracy. This mainly affects the computed bias ratio (*β*), but not so much temporal dynamics as measured by the computed correlation (*r*). In the calibration period the correlation is low ($r = 0.74$) because operation rules imposed on the model reflect the current situation (as effective during the 2000s), whereas the actual historic operation of Kariba and Cahora Bassa reservoirs changed over time (see discussion in previous section). In contrast to the calibration period, the correlation between simulated and observed discharge is high ($r = 0.95$) in the independent evaluation period, with observed data at Tete available from 1952 to 1960. The first seven years represent undisturbed (pristine) conditions, whereas the last two years are affected by the filling of Kariba reservoir.

Of greater interest than the poor bias ratio and correlation at Tete is the model performance for simulation of Zambezi discharge at Victoria Falls. Discharge data measured at this gauge are considered to be accurate – and are not affected by upstream reservoir operations. Model performance during the calibration period is high, but the bias ratio becomes significant ($\beta = 1.19$) in the evaluation period. The most likely source for this bias is that the precipitation inputs are already biased. From the calibration to the evaluation periods mean annual precipitation increased by +3%, but observed discharge decreased by −4%. Even though these are small changes, it is counter-intuitive that discharge decreases when precipitation increases. Here, the low density of precipitation stations has to be considered in the upper Zambezi basin, which is on average approximately one station per 21,000 km^2 in the calibration period, but even lower during the evaluation period (see Fig. 2). An under-estimation of discharge in the evaluation period is also obtained at the upstream gauge Lukulu, albeit the period with available data is only 7 years.

The under-estimation of Kafue River discharge at the gauge Kafue Hook Bridge during the calibration period is the result of a large negative bias (−34%) during a 5-year period (1978–1982), which coincides with the start of operation of nearby Itezhitezhi reservoir. The source of this bias is not clear, but it could be related to the accuracy of the precipitation data or the discharge data. Outside this 5-year period the simulation shows only a small bias – this also applies to the independent evaluation period.

Scenario Simulation

The calibrated model was applied for simulation of a number of pre-defined scenarios (see Table 3). The scenario simulations are always compared against the "Baseline" scenario representing current water resources management (reservoirs, operation rules, irrigation withdrawals) in the basin but using historic climate of the period 1961–1990. The analysis focuses on Zambezi River discharge at Tete in Mozambique.
Table 5 lists mean annual scenario results. Mean annual discharge in the Baseline scenario amounts to approximately 2600 m^3/s, with values ranging from around 1750 m^3/s to 3700 m^3/s in the scenario simulations.

Table 5. Mean annual scenario simulation results for the Zambezi basin upstream of Tete.

Scenario	[m^3/s]	[%]	Evaporation [m^3/s]	[m^3/s]
	Discharge	ΔQ to baseline	Wetlands	Diversions
Historic scenarios				
Calibration	2514	−3	311	15
Evaluation	3042	17	333	15
Reference scenario				
Baseline	2597	–	311	74
Water resources development scenarios				
Pristine	2956	14	311	0
Moderate development	2498	−4	310	179
High development	2137	−18	308	564
Climate model scenarios				
CNRM near future	2846	10	343	75
MPI near future	2241	−14	289	73
CNRM far future	2965	14	324	76
MPI far future	2127	−18	267	70
Climate sensitivity scenarios				
P −10%	1754	−32	243	71
P +10%	3725	43	364	76
T +2 °C	2375	−9	304	73
T +4 °C	2176	−16	295	72

Total evaporation losses from reservoirs amount to 437 m^3/s in the Baseline scenario. This value ranges from 418 to 499 m^3/s in the other scenarios. The differences are caused by:

- Different number of reservoirs (Batoka Gorge and Mphanda Nkuwa are included in the Moderate and High development scenarios).
- Increase in temperature causes higher evaporation rates.
- Variations in inflow (due to changes in precipitation, evapotranspiration, diversions) affect reservoir water levels. For example, lower inflows can result in lower water levels, thereby decreasing evaporation losses due to smaller surface area of the water body.

More than 90% of the total reservoir evaporation occurs from Kariba and Cahora Bassa reservoirs. These are significant losses of water and the main reason that under the Pristine scenario (with no reservoirs) discharge is considerably larger than in the other scenarios.

In addition to the reservoirs, water also evaporates from the natural wetlands and floodplains – with mean annual evaporation losses ranging from 243 to 364 m^3/s between the scenarios. The contribution to total evaporation from the individual wetlands is roughly 40% from Kafue Flats, 25% from Barotse Floodplain, 25% from Chobe Swamps, and 10% from Kwando Floodplain. The sum of evaporation losses from reservoirs and wetlands amounts to approximately 750 m^3/s, which is more than 20% of the runoff that enters the river network.

Diversions amount to 74 m^3/s in the Baseline scenario, which is small compared to the evaporation losses from reservoirs and wetlands. However, diversions increase to 179 m^3/s in the Moderate development scenario, and to 564 m^3/s in the High development scenario. This means that irrigation levels under the High development scenario have a similar magnitude as evaporation losses that are already occurring from existing reservoirs. Under this scenario mean annual discharge decreases by −18% as compared to the Baseline scenario. 87% of the irrigation demand (Table 3) can be met by the simulated diversions (Table 5). Similar percentages are obtained in the Moderate development and Baseline scenarios – albeit with much lower diversion amounts. Shortages for meeting irrigation demand occur when reservoir water levels fall below minimum operation levels. This situation occurs at Zimbabwean tributaries under all scenarios, but also in dry years at Kariba reservoir under the High development scenario.

It is clear that an implementation of irrigation projects will cause a decrease in discharge due to increased diversions. The impact of future climate is less clear, though. Contrasting results are obtained for the scenarios based on climate data of GCMs. For the near future (2021–2050) the scenario based on CNRM climate data projects an increase in discharge of +10%, whereas MPI projects a decrease of −14%. These differences are even larger for the far future (2071–2100), with projected changes of +14% versus −18%.

To disentangle the effects of changes in precipitation and temperature the last four scenarios listed in Table 5 present assessments for changes superimposed on historic climate (delta-change approach). If temperature increases by +4 °C then discharge decreases by −16%. An even larger decrease in discharge of −32% is obtained for a reduction of precipitation by −10%. An increase in precipitation by +10% results in an increase of discharge by +43%.

The percentage changes in mean annual discharge are not evenly distributed during a year, as evident in an analysis of seasonality in discharge (Fig. 10, top left). By far the largest differences to the Baseline scenario are obtained with the Pristine scenario, with a more pronounced seasonality. The main reason is that the Pristine scenario does not include any reservoirs. The reservoir operation results in a strong attenuation of the seasonal flood peak and an increase of discharge during the dry period. This is even clearer when analysing the distribution of flows (Fig. 10, top right). In the Pristine scenario high flows are increased, but low flows are much lower, even though the mean annual discharge is larger.

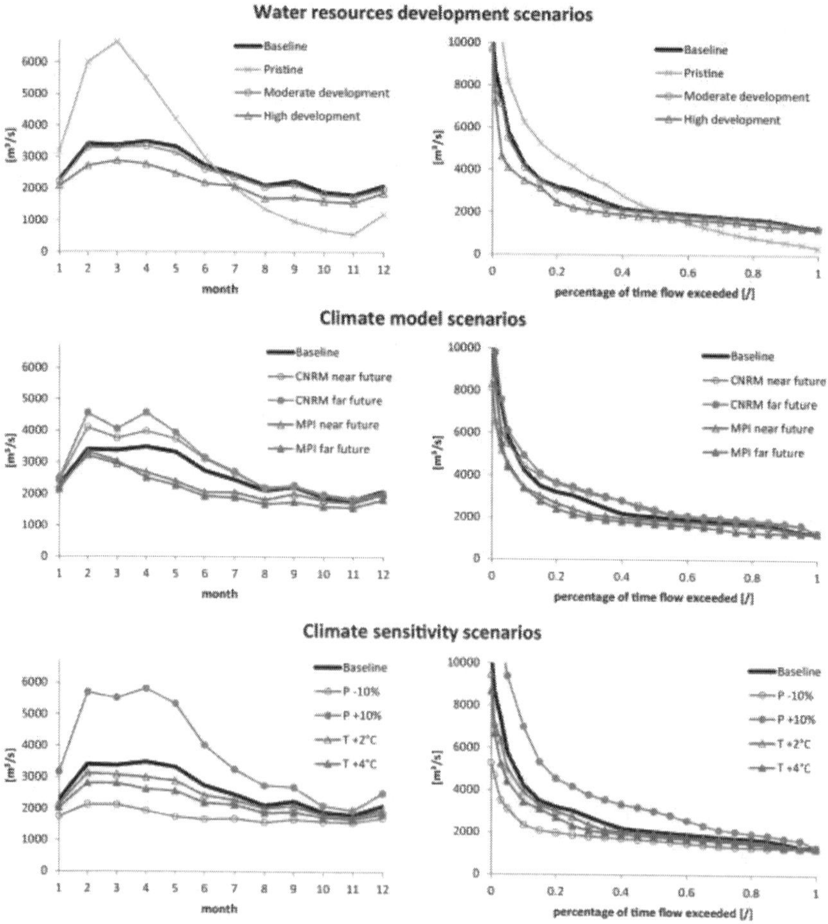

Figure 10. Simulated discharge conditions of Zambezi River at Tete under various scenarios. Top: water resources development scenarios. Middle: climate scenarios. Bottom: climate sensitivity tests. Left: seasonality in discharge. Right: monthly flow duration curve.

For the High development scenario the magnitude of changes in seasonality and distribution of discharge are considerably smaller than for the Pristine scenario (Fig. 10, top) – and the changes are insignificant for the Moderate development scenario. In the High development scenario a relatively constant decrease is obtained for the seasonality in discharge (Fig. 10, top left), which is the result of the interplay of seasonality in irrigation demand and reservoir operation. For the distribution of flows (Fig. 10, top right) there are significant decreases for higher flows, but almost no decreases for low flows. This is caused by constant releases of reservoirs during dry periods.

Fig. 10 (middle) shows the changes in seasonality and distribution of discharge in the scenarios based on future projections of climate models. The differences between the climate models are large, whereas the time period (near versus far future) is of limited importance. This reflects the lower sensitivity to temperature – which is different in the two time periods – and the higher sensitivity to precipitation – which is different in the two climate models. For the far future scenario with MPI climate data the low flows decrease more than in other scenarios. This is caused by lack of precipitation, which cannot be fully compensated by reservoir operation during dry periods.

The results for the climate sensitivity scenarios are shown in Fig. 10 (bottom). In the scenario with +10% increase in precipitation there is a pronounced seasonality in discharge, whereas for −10% decrease in precipitation seasonality almost completely disappears (Fig. 10, bottom left). For this scenario, 90% of the time discharge is almost constant at approximately 2000 m^3/s (Fig. 10, bottom right).

The monthly flow duration curves shown in Fig. 10 suggest that there will not be severe changes for low flows in the future. As Fig. 11 shows, annual discharge of individual years will also not change significantly in the future for the driest years. Interestingly, the lowest annual discharge was simulated for the Pristine scenario, with no reservoirs to sustain minimum flow in very dry periods. In contrast, there are significant differences in the annual discharge in the wettest years. The scenarios based on climate model data project that the highest annual discharge will be significantly larger in the far future than in the near future. These changes are independent from the changes in mean annual discharge. However, any interpretation of extreme events based on climate model data should be cautious (Kundzewicz and Stakhiv, 2010, Wilby, 2010 and Blöschl and Montanari, 2010).

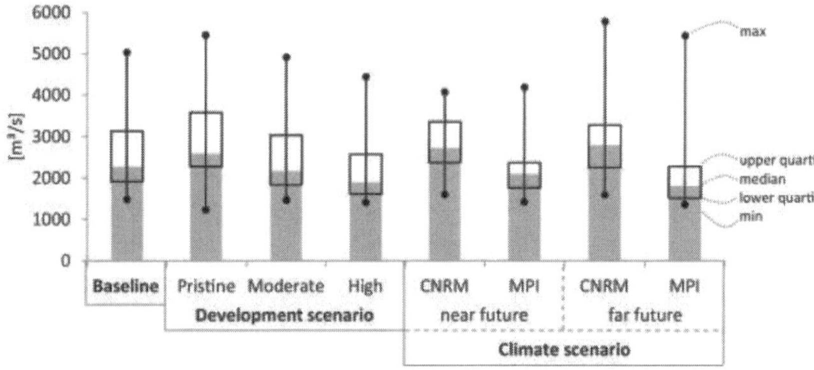

Figure 11. Distribution of annual discharge. Box-plots show the annual discharge of individual years simulated under various scenarios. Zambezi River at Tete simulated for 30-year periods. Comparison of baseline scenario to water resources development scenarios and climate scenarios.

DISCUSSION

In this section we discuss the simulation results and also give a brief overview about possible sources of uncertainties in the impact modelling.

Simulation Results

The model simulations obtained for historic conditions are consistent with available observations. This applies for a visual comparison of simulated and observed discharge and reservoir water level data, as well as performance statistics in the calibration and independent evaluation periods.

The performance statistics are higher than in previous studies. For example, for the discharge simulation at Victoria Falls during calibration period Harrison and Whittington (2002) obtained a correlation coefficient R^2 of 0.61, which is lower than the results presented here with R^2 of 0.88. Similarly, Winsemius et al. (2006) report for their two models a Nash–Sutcliffe Efficiency NSE of 0.72 and 0.82 respectively, whereas we obtained a slightly higher performance with NSE of 0.88. Note that Winsemius et al. did only apply their model to the upper Zambezi and did not focus on impact modelling. Unfortunately, the other impact modelling studies of the whole Zambezi basin (Hoekstra,

2003, Yamba et al., 2011 and Beck and Bernauer, 2011) do not report performance statistics. However, we believe that the model simulations presented here are among the most accurate – if not best – models for simulation of Zambezi discharge currently available.

The exact reason for the higher model performance as compared to previous studies remains unclear. It may be related to improved input data (GPCC), calibration method, consideration of wetlands and river routing. The latter two are important for simulation of timing of Zambezi discharge (Cohen-Liechti et al., 2014) and would cause serious modelling problems if not explicitly considered, with the risk of corrupting parameter values to obtain simulations that are "right for the wrong reason" (Refsgaard and Henriksen, 2004). The higher performance is most likely not related to the structure of the water balance model (see Fig. 4, left), as here the applied models are all very similar in the various studies.

The evaluation of historic discharge conditions (see Fig. 5 and Fig. 6) also shows the considerable impact of the large reservoirs and the problem of reservoir operation; where (ad hoc?) release decisions at upstream reservoirs complicate simulation of downstream discharge. Different sets of operation rules would have to be applied to different time periods, but instead fixed operation rules – as effective during the 2000s – were imposed on the model. Therefore, simulations in the downstream sections (e.g. at Tete) frequently show deviations to observations.

Due to the above mentioned peculiarities of Zambezi discharge in downstream sections, we focussed on the simulation results averaged over the land-surface – thereby excluding the confounding impacts of reservoirs – to learn more about the hydrology in the context of the seasonal water balance (see Fig. 9). The hydrology in the Zambezi basin is characterized by representing a water limited system – as opposed to energy limited. Already under historic climate the potential evapotranspiration cannot be met by the actual evapotranspiration (see Fig. 9), simply because there is not enough water stored in the soil due to insufficient annual precipitation amounts. Therefore, any increases in temperature – and consequently increases in potential evapotranspiration – have a small impact on discharge. In contrast, small changes in precipitation have large impacts on discharge. This was already observed in the past, where discharge is considerably larger in wet years than in dry years and the model simulations are well in line with this observation (see Fig. 8). Under such conditions any projections with climate models have to be interpreted with caution – only small variations (increases/decreases) in precipitation projections cause large differences in the impact on discharge. This was also confirmed by the sensitivity tests (see Table 5 and Fig. 10, bottom) – where a decrease of precipitation by −10% caused a decrease in discharge by almost −850 m^3/s, or −32%. Note that this high sensitivity of discharge to precipitation contrasts the conclusions of Beck and Bernauer (2011) that climate has relatively small effects on water availability in the Zambezi basin, which may be related to their approach of calibration to long-term average conditions.

Our simulations under climate change scenarios show a range of −14% to +10% for mean annual Zambezi discharge at Tete in the near future (2021–2050 as compared to Baseline simulation 1961–1990). These results (and the large uncertainty) have to be interpreted within the context of the results of previous studies. Harrison and Whittington (2002) focussed on the upper Zambezi River at Victoria Falls. For the 2080s their three climate scenarios show a warming of about +5 °C and a reduction in rainfall between −2% and −18%, which results in a reduction in runoff by −10% to −36%. In a preliminary analysis the World Bank (2010) used GCM data (A1B emission scenario) for the whole Zambezi region. For 2030 they estimate a change in runoff between −13% and −34% (depending on the sub-region). Beilfuss (2012) summarized existing climate change assessments for the Zambezi and concludes that by 2050 runoff is likely to decrease by −26% to −40% if the reduction in rainfall lies between −10% and −15%. This corresponds well to our climate sensitivity tests where for a reduction of −10% in rainfall the simulation shows a reduction of −32% in discharge. However, apart from these dramatic projections with reduction in flows we also have to acknowledge that rainfall may actually increase in the future, highlighting the uncertainty in the climate model scenarios.

In addition to climate change, also future development of large-scale irrigation is expected to have a considerable impact on Zambezi discharge. For the high-level irrigation development the simulations show a decrease of mean annual Zambezi discharge at Tete by −460 m^3/s (−18%). This is similar in magnitude as the reduction caused by evaporation from existing reservoirs (437 m^3/s). Overall, the impact of the existing reservoirs is much larger than just reducing mean annual discharge, because in addition they also affect the discharge conditions. The construction of Kariba and Cahora Bassa dams and the reservoir operation policies have led to a strong alteration of the natural seasonality and distribution of discharge (flow duration curve). This became evident in a comparison between Baseline and Pristine scenarios (see Fig. 10). No such significant changes were found in the other analysed scenarios representing possible future conditions. This means that – and we believe this is a significant finding –

the biggest changes for Zambezi discharge have already occurred in the past.

Apart from the Pristine scenario, in all other scenarios studied, no pronounced changes were obtained for neither monthly low flows (see monthly flow duration curves in Fig. 10) nor annual discharge in the overall driest years (see Fig. 11). The reason is that Kariba and Cahora Bassa reservoirs are sufficiently large to support low flows in dry periods by drawing down the water levels. However, if more extreme (i.e. drier) climate scenarios were included, then the reservoirs would reach their minimum operation levels and discharge would drastically decrease in dry years.

The impact of the reservoirs becomes larger for scenarios with drier conditions. For example, if precipitation decreased by −10%, this would result in almost constant flows without any seasonal fluctuations (Fig. 10, bottom). This would have dramatic consequences for downstream ecology. Under such conditions reservoir operation rules should be refined to impose seasonal fluctuations on the reservoir releases (Beilfuss, 2010).
This large impact of the reservoir operation enables water resources managers to actively control the downstream discharge conditions. Poor planning or lack of co-operation obviously can lead to negative impacts, but on the other hand good planning can have many positive impacts. Therefore, balanced solutions are required considering flood safety, hydropower generation, irrigated agriculture and ecological aspects.

Sources of Uncertainties

The hydrological impact modelling in this study is affected by several uncertainties. Exact quantification of these uncertainties would significantly increase the scope of this study and is left for future work. However, it is still worthwhile to discuss where these uncertainties may arise from for the hydrological model and future scenarios.

The main sources of uncertainty for the hydrological model set-up are listed below:

- Observed discharge data: Measurement errors due to inaccurate rating curves.
- Historic climate data: Poor precipitation station density.
- Water balance model: Model structure, model parameters.
- Water allocation model: Reservoir operation rules.

Of the uncertainties listed above it is deemed that the observed discharge data are most important. As the model is calibrated to closely match these data, any systematic biases in the observed data would also affect the simulations. Before calibration, plausibility checks (double-mass plots, upstream–downstream comparisons) resulted in rejection of discharge data from a number of gauges, to avoid an over-fitting of the model to biased data. However, also the remaining gauges may be – and most likely are – affected by biases, affecting computation of mean flows, but not so much the temporal dynamics of flows. In summary, even though the accuracy of discharge observations is to some degree questionable, the data of the used gauges are the best information available.

For the historic climate data the poor precipitation station density is a concern especially in the upper Zambezi basin – with approximately one station per 21,000 km^2. The station density is highest – and uncertainty is lowest – during the period 1961–1990, which was also used for calibration and for the Baseline scenario. The used precipitation data set (GPCC) is currently the best available long-term, observational data set in the Zambezi basin. The number of stations included is almost twice as high as in the well-known data set of CRU. Other interesting data sources would include satellite-based data such as TRMM (Tropical Rainfall Measurement Mission of NASA, Huffman et al., 2007), albeit TRMM data are only available since 1998. A comparison of these data-sets could be an attempt to quantify the uncertainty in the historic precipitation model inputs, but faces the obstacle of lack of overlapping time-period with good quality ground-based data (Cohen-Liechti et al., 2012).

Uncertainties in model structure and parameters have received considerable attention in the scientific literature, and there are also a few examples of such studies in southern Africa (e.g. Winsemius et al., 2006, Winsemius et al., 2009 and Hughes et al., 2010). These studies give interesting insights into model behaviour and performance of alternative models. However, we believe that a well calibrated model, with high performance and thorough evaluation – including for example separate evaluation in wet and dry years – increases the confidence also for simulation under various scenarios. An important assumption here is that parameter values obtained from calibration to historic conditions are also applicable for simulation under future conditions, thereby ignoring possible impacts of land-use change and dependence of calibrated model parameters on climate characteristics (Singh et al., 2013). An inter-comparison study – juxtaposing results of different modelling approaches – would be required to quantify the hydrological model uncertainty.

Simulations under future development and climate scenarios strictly have to be interpreted as What-if analyses, as opposed to deterministic forecasts. No likelihoods are attached to these scenarios. Future development of irrigation and dam projects in the basin depends on political decisions, economic development, population growth, and sound water resources planning. Climate model projections are affected by emission scenarios, natural climate variability, climate model errors, downscaling technique and bias correction. All these aspects result in a large range of uncertainty. Within the scenarios, there are different sensitivities of the results. For the development scenarios, the impact of future irrigation projects is more important than future dam projects. For the climate scenarios, higher temperatures cause gradual decrease in discharge, but results are more (highly) sensitive to precipitation. Here, the two climate models used do not agree on the sign in the change of future precipitation.

This uncertainty in future precipitation is the most important source of uncertainty for future Zambezi discharge. As a logical next step, the analysis should be expanded by using a whole ensemble of climate models, as shown, e.g. by Kling et al. (2012) for the upper Danube basin. Ideally, the climate data should be based on regional climate models (RCMs) that are currently applied in on-going research projects for the African continent. RCMs have a much finer spatial resolution and are deemed to be superior to GCM projections (as used in this study), especially regarding the simulation of the seasonal shift of the Inter-Tropical Convergence Zone (ITCZ), which controls precipitation.

Table 6 lists a first analysis of climate projections for the Zambezi basin simulated by three RCMs in the recently finished ENSEMBLES project (Paeth et al., 2011). All three analysed RCMs project a decrease in precipitation for the Zambezi basin – with projections for 2071–2100 of −9% by INM and −18% by ICTP. These decreases are significantly larger than the decrease in the analysed GCM data of this study – with a maximum decrease of −5% projected by MPI for 2071–2100 (see Table 1). Decreases in precipitation by −10% and more would have dramatic impacts on discharge in the Zambezi River, where from the sensitivity analyses presented here it is expected that annual discharge would decrease by more than −30% (see Table 5). Therefore, we recommend focusing future work on assessing the impact of an ensemble of regional climate model projections, which will be made available via the Coordinated Regional Climate Downscaling Experiment for Africa (CORDEX-Africa, see e.g. Nikulin and Jones, 2011 and Kalognomou et al., 2013).

Table 6. Climate model data of the ENSEMBLES project for the Zambezi catchment upstream of Tete. A1B emission scenario. Spatial resolution of RCMs is approximately half degree. Raw (uncorrected) RCM data.*P*: mean annual precipitation, *T*: mean annual air temperature, ΔP: change in precipitation relative to 1990–2010, ΔT: change in air temperature relative to 1990–2010.

Institute	GCM/RCM	Period	*P* [mm/a]	*T* [°C]	ΔP [%]	ΔT [°C]
DMI[a]	ECHAM5-r3/HIRHAM5	1990–2010	770	24.3	–	–
		2021–2050	724	25.5	−6	1.2
		2071–2100	n/a	n/a	n/a	n/a
ICTP[b]	ECHAM5-r3/RegCM3	1990–2010	808	22.3	–	–
		2021–2050	780	23.4	−3	1.1
		2071–2100	660	26.7	−18	4.4
INM[c]	HadCM3Q0/RCA3	1990–2010	994	22	–	–
		2021–2050	989	23.4	−1	1.4
		2071–2100	901	25.5	−9	3.5

[a] DMI: Danmarks Meteorologiske Institut, Denmark.
[b] ICTP: International Centre for Theoretical Physics, Italy.
[c] INM: Instituto Nacional de Meteorología, Spain.

OUTLOOK

This study is embedded in a broad scale initiative to assess – and prepare for – climate change impacts in Mozambique (INGC, 2009). The modelling tools and databases of this study have been implemented in a web-based, interactive Decision Support System (DSS, online access at http://zdss.ingc.gov.mz/[1]). Thereby, the whole database used in this study is readily available to the general public. In addition to data export, the DSS allows editing and creating development and climate scenarios, as well as inserting computation points to query discharge simulations at points of interest along the river network. Mozambican analysts have been trained on the DSS, such that further work can focus on:

- Studies for individual Mozambican tributaries of the Zambezi.
- Adapting operation rules of existing large reservoirs.

- Impact assessment of planned water resources projects including dams, irrigation projects as well as inter-basin transfers (e.g. upper Zambezi water diversion to Botswana/South Africa).
- Assessments for new climate scenarios (CORDEX).
- Combined assessment of climate and development scenarios.
- Teaching of students at university.

In a recent update, the DSS has been extended to include simulation of energy generation at hydro-power plants, discharge simulation in daily time-steps, and coupling with flood mapping in the lower reaches of the Zambezi. The training on – and the work with – the DSS is one building block for capacity increase in Mozambique. Thereby, this will foster awareness and preparedness for possible future impacts of water resources development and climate change in the Zambezi basin.

CONCLUSIONS

This study focussed on the hydrological impact modelling of water resources development and climate change scenarios on discharge conditions in the Zambezi basin. A river basin model was calibrated with historic data, before being applied for a number of scenarios.

A specific objective of this study was a thorough evaluation of the model simulations, as there has been a lack thereof in previous impact assessment studies. Our simulations of historic conditions are consistent with available observations. This applies for simulation of river discharge as well as reservoir water levels. The model performance statistics do not drop significantly when moving from the calibration period to an independent evaluation period. Overall, the performance statistics are superior to previous studies. The accurate discharge simulations thereby increase the confidence in the impact assessment.

The simulation of historic conditions enables the following conclusions:

- There are large inter-annual variations in discharge. Discharge in wet years is more than twice as large as discharge in dry years, which is related to small variations in precipitation. This high sensitivity of discharge to precipitation was not fully appraised in previous impact modelling studies.
- Runoff is only a small component of the water balance and shows a strong seasonality due to distinctive rainy/dry seasons. Most of the

rainfall is lost via evapotranspiration, with the actual evapotranspiration rates considerably lower than the potential ones throughout most of the year. Here, changes in soil moisture play a key role for the seasonal water balance.
- Natural wetlands and operation of large reservoirs have a large impact on discharge. This has to be explicitly included when building hydrological models of the Zambezi River basin.

Several scenarios were defined considering future developments for irrigation withdrawals and dams as well as climate change scenarios, with the following main findings:

- The biggest changes in the Zambezi basin have already occurred in the past. The construction of large reservoirs caused a decrease in discharge by evaporation and significantly altered the discharge conditions by reservoir operation. Low flows have been increased and high flows decreased.
- The simulated decrease of −18% in mean annual discharge due to possible future high-level irrigation development is of similar magnitude as the total evaporation from existing reservoirs. Moderate irrigation development and planned new dams have only minor impacts on Zambezi River discharge. The projected irrigation demand in both future scenarios can be met to a similar degree (almost 90%) as the current demand (when climate change is not considered).
- Future climate causes large uncertainties in future discharge. This is caused by the high sensitivity of discharge to precipitation, but the analysed climate models do not agree on the sign of future changes in precipitation.
- Even though mean discharge is strongly affected by any changes in precipitation, the low flows remain almost unchanged due to constant releases from reservoirs during dry periods. Low flows may be drastically reduced if analysing scenarios with more than −10% change in future precipitation.
- Future warming will cause higher actual evapotranspiration and therefore lower discharge. A warming of +4 °C has a similar impact as high-level irrigation development.

These scenarios show that the impact on future Zambezi River discharge can be quite large. At the same time, the human-induced changes in the past may have been larger than the changes in the future. This also means that human management – if adapted well to the changing conditions – can contribute substantially to mitigating negative effects of a changing climate. Here, the largest uncertainty relates to future precipitation.

Current, on-going research efforts with regional climate models applied to Africa should enable more detailed assessments within an ensemble modelling framework.

ACKNOWLEDGEMENTS

This project was carried out in collaboration with HYDROC, Germany for the National Institute of Disaster Management (INGC), Mozambique. Funding was provided from the United Nations Development Programme (UNDP). Many thanks go to the various institutions supporting this project, but most notably to the Physics Department at Eduardo Mondlane University (UEM) and the Centro Naçional Operativo de Emergência (CENOE), both located in Maputo, Mozambique. Historic data were supplied from the Direcção Naçional das Águas in Mozambique, Zambezi River Authority in Zambia, Global Runoff Data Centre in Germany, Global Precipitation Climatology Centre in Germany, and the Climatic Research Unit in the UK. Climate model data were provided from the EU WATCH project and the EU ENSEMBLES project. We thank for the timely and constructive comments of two anonymous reviewers and the editor (Denis Hughes) at Journal of Hydrology: Regional Studies. We also thank for the comments of two anonymous reviewers, the associate editor (Harry Lins) as well as the editor (Demetris Koutsoyiannis) of Hydrological Sciences Journal, where this paper was originally submitted in 2012.

APPENDIX A

When a fine temporal discretization (e.g. $\Delta t = 24$ h) is used, then the water balance model can be formulated with the simple equations given below. To overcome numerical artefacts that arise with such a discrete numerical formulation, the actual model code uses (1) an internal adaptive temporal discretization for the soil module and (2) the analytical solution of the differential equation that gives the exact result for $\Delta t \rightarrow 0$ for the base flow module. The interception module uses daily time-steps with alternating wet and dry days for precipitation within a month. The parameters of the model are listed inTable 7 and the variables are listed in Table 8. Δt is in the units of h. The index t identifies the time-step and the index i identifies the month (January–December). To improve readability, the index for identifying the sub-basin is omitted for all parameters and variables. Each sub-basin is treated lumped (i.e. no spatial variability within sub-basin).

Table 7. Parameters of the water balance model. Range specifies the parameter bounds during calibration. A single value is given if the parameter is not calibrated.

Parameter	Units	Range	Description
mETP	mm/d	–	Long-term mean monthly potential evapotranspiration in reference period.
mT	°C	–	Long-term mean monthly air temperature in reference period.
F	$°C^{-1}$	0.025	Empirical factor obtained from sensitivity tests with Penman-Monteith.
TF	–	0.5	Through-fall fraction for interception of precipitation.
SImax	mm	2	Interception storage capacity.
S1max	mm	100–2000	Soil storage capacity.
S1crit	–	0.5	Critical soil moisture below which evapotranspiration is reduced.
Beta	–	1–10	Soil exponent for runoff generation.
K2	–	0–1	Fraction for separation of surface flow.
K3	h	1000	Base flow recession coefficient.

Table 8. Variables of the water balance model. Flux variables represent sums over the time-step. State variables give the water storage at the end of the time-step.

Variable	Units	Type	Description
P	mm	Input	Precipitation sum over time-step.
T	°C	Input	Mean air temperature within time-step.
ETP	mm	Flux	Potential evapotranspiration.
ETPI	mm	Flux	Available potential evapotranspiration for interception.
ETAI	mm	Flux	Actual evaporation from interception storage.
ETPR	mm	Flux	Remaining potential evapotranspiration after interception.
SI	mm	State	Water stored in interception storage.
PNET	mm	Flux	Net-precipitation after interception.
S1	mm	State	Water stored in soil storage.
ETAG	mm	Flux	Actual evapotranspiration from soil storage.
ETA	mm	Flux	Total actual evapotranspiration.
Q0	mm	Flux	Runoff generation.
Q1	mm	Flux	Surface flow.
Q2	mm	Flux	Percolation to base flow storage.
S3	mm	State	Water stored in base flow storage.
Q3	mm	Flux	Base flow.
QSIM	mm	Flux	Runoff.
X1 to *X3*	mm	–	Temporary variables for interception module.

Potential evapotranspiration

$$ETP_t = mETP_i \cdot [(T_t - mT_i) \cdot F + 1] \cdot \Delta t/24$$

Interception module

$$X1 = SI_{t-1} + (1 - TF) \cdot Pt$$

$$X2 = \max\{0, X1 - SI_{max}\}$$

$$X3 = X1 - X2$$

$$ETPI_t = ETP_t \cdot \frac{1.5 \cdot SI_{max}}{1.5 \cdot SI_{max} + 1}$$

$$ETAIt = \min\{ETPIt, X3\}$$

$$ETPRt = ETPt - ETAIt$$

$$SIt = X3 - ETAIt$$

$$PNETt = X2 + TF \cdot Pt$$

Soil module

if $S1_{t-1} < S1_{crit} \cdot S1_{\max}$:

$$ETAG_t = ETPR_t \cdot S1_{t-1}/(S1_{crit} \cdot S1_{\max})$$

$$\text{if } S1_{t-1} \geq S1_{crit} \cdot S1_{max}:$$

$$ETAG_t = ETPR_t$$

$$ETA_t = ETAI_t + ETAG_t$$

$$Q0_t = PNET_t \cdot (S1_{t-1}/S1_{max})^{Beta}$$

$$S_t{}_1 = S_{1t-1} + PNET_t - ETAG_t - Q_t{}_0$$

Surface flow module

$$Q_t{}_1 = K2 \cdot Q_t{}_0$$

$$Q_t{}_2 = (1 - K2) \cdot Q_t{}_0$$

Base flow module

$$Q3_t = \Delta t \cdot S3_{t-1}/K3$$

$$S_t{}_3 = S_{3t-1} + Q_t{}_2 - Q_t{}_3$$

Runoff

$$QSIM_t = Q_t{}_1 + Q_t{}_3$$

REFERENCES

1. Allen, R., Pereira, L., Raes, D., 1998. Crop Evapotranspiration – Guidelines for Computing Crop Water Requirements – FAO Irrigation and Drainage Paper 56. Food and Agriculture Organization of the United Nations (FAO).
2. Beck, L., Bernauer, T., 2011. How will combined changes in water demand and climate affect water availability in the Zambezi river basin? Global Environ. Change 21 (3), 1061–1072.
3. Beilfuss, R., dos Santos, D., 2001. Patterns of Hydrological Change in the Zambezi Delta, Mozambique. Working Paper #2, Program for the Sustainable Management of Cahora Bassa Dam and the Lower Zambezi Valley, Report, 159 pp.
4. Beilfuss, R., 2010. Modelling trade-offs between hydropower generation and environmental flow scenarios: a case study of the Lower Zambezi River Basin, Mozambique. Int. J. River Basin Manage. 8, 331–347.
5. Beilfuss, R., 2012. A Risky Climate for Southern African Hydro. International Rivers, Berkeley, CA, USA, pp. 60. Bergström, S., 1995. The HBV model. In: Singh, V.P. (Ed.), Computer Models of Watershed Hydrology. Water Resources Publications, Highlands Ranch, CO, USA, ISBN: 0-918334-91-8.
6. Blöschl, G., Montanari, A., 2010. Climate change impacts – throwing the dice? Hydrol. Process. 24, 374–381.
7. Cohen-Liechti, T.,Matos,J.P.,Boillat,J.L., Schleiss,A.J., 2012 Comparisonand evaluationof satellite derived precipitationproducts for hydrological modeling of the Zambezi River basin. Hydrol. Earth Syst. Sci. 16, 489–500.
8. Cohen-Liechti, T., Matos, J.P., Ferràs-Segura, D., Boillat, J.L., Schleiss, A.J., 2014. Hydrological modelling of the Zambezi River Basin taking into account floodplain behaviour by a modified reservoir approach. Int. J. River Basin Manage. 12 (1), 29–41.
9. Duan, Q., Sorooshian, S., Gupta, V., 1992. Effective and efficient global optimization for conceptual rainfall-runoff models. Water Resour. Res. 28 (4), 1015–1031.
10. Gandolfi, C., Salewicz, K.A., 1991. Water resources management in the Zambezi Valley: analysis of the Kariba operation. In: IAHS Publ. No. 201, Proceedings of the Vienna Symposium, August 1991.
11. Gupta, H.V., Kling, H., Yilmaz, K.K., Martinez, G.F., 2009. Decomposition of the mean squared error and NSE performance criteria: implications for improving hydrological modelling. J. Hydrol. 377, 80–91.
12. Harrison, G.P., Whittington, H.W., 2002. Susceptibility of the Batoka Gorge hydroelectric scheme to climate change. J. Hydrol. 264, 230–241.

13. Hoekstra, A.Y., 2003. Water scarcity in the Zambezi basin in the long-term future: a risk assessment. Integr. Assess. 4 (3), 185–204.
14. Huffman, G.J., Adler, R.F., Bolvin, D.T., Gu, G., Nelkin, E.J., Bowman, K.P., Hong, Y., Stocker, E.F., Wolff, D.B., 2007. The TRMM Multisatellite Precipitation Analysis (TMPA): quasiglobal, multiyear, combined-sensor precipitation estimates at fine scales. J. Hydrometeor. 8, 38–55.
15. Hughes, D.A., Kapangaziwiri, E., Sawunyama, T., 2010. Hydrological model uncertainty assessment in southern Africa. J. Hydrol. (387), 221–232.
16. INGC, 2009. In: Asante, K., Brundrit, G., Epstein, P., Fernandes, A., Marques, M.R., Mavume, A., Metzger, M., Patt, A., Queface, A., Sanchez del Valle, R., Tadross, M., Brito, R. (Eds.), Main Report: INGC Climate Change Report: Study on the Impact of Climate Change on Disaster Risk in Mozambique. INGC, Mozambique, p. 338.
17. IPCC, 2007. Summary for Policymakers. In: Solomon, S., Qin, D., Manning, M., Chen, Z., Marquis, M., Averyt, K.B., Tignor, M., Miller, H.L. (Eds.), Climate Change 2007: The Physical Science Basis. Contribution of Working Group I to the Fourth Assessment Report of the Intergovernmental Panel on Climate Change. Cambridge University Press, UK, pp. 1–18.
18. Kalognomou, E.A., Lennard, C., Shongwe, M., Pinto, I., Favre, A., Kent, M., Hewitson, B., Dosio, A., Nikulin, G., Panitz, H.J., Büchner, M., 2013. A diagnostic evaluation of precipitation in CORDEX models over southern Africa. J. Climate 26, 9477–9506.
19. Kling, H., Fuchs,M., Paulin,M., 2012.Runoff conditions in the upper Danube basin under an ensemble of climate change scenarios. J. Hydrol. 424–425, 264–277.
20. Koutsoyiannis, D., 2011. Hurst–Kolmogorov dynamics and uncertainty. J. Am. Water Resour. Assoc. 47 (3), 481–495.
21. Kundzewicz, Z.W., Stakhiv, E.Z., 2010. Are climate models "ready for prime time" in water resources management applications, or is more research needed? Hydrol. Sci. J. 55 (7), 1085–1089. Lehner, B., Verdin, K., Jarvis, A., 2008. New global hydrography derived from spaceborne elevation data. Eos Trans., AGU 89 (10), 93–94.
22. Lehner, B., Liermann, C.R., Revenga, C., Vörösmarty, C., Fekete, B., Crouzet, P., Döll, P., Endejan, M., Frenken, K., Magome, J., Nilsson, C., Robertson, J.C., Rödel, R., Sindorf, N., Wisser, D., 2011. High-resolution mapping of the world's reservoirs and dams for sustainable river-flow management. Front. Ecol. Environ. 9 (9), 494–502.
23. Lenderink, G., Buishand, A., van Deursen, W., 2007. Estimates of future discharges of the river Rhine using two scenario methodologies: direct versus delta change approach. Hydrol. Earth Syst. Sci. 11 (3), 1145–1159.

24. Lins, H.F., Cohn, T.A., 2011. Stationarity: wanted dead or alive? J. Am. Water Resour. Assoc. 47 (3), 475–480. Matalas, N.C., 2012. Comment on the announced death of stationarity. J. Water Resour. Plann. Manage. 2012, 311.
25. Matos,J.P., Cohen, T., Boillat,J.L., Schleiss,A.J., Portela,M.M., 2010.Analysis offlow regime changes due to operation oflarge reservoirs on the Zambezi River. In: Christodoulou, Stamou (Eds.), 6th International Symposium on Environmental Hydraulics. Taylor & Francis Group, London, pp. 337–342, ISBN 978-0-415-58475-3.
26. Mazvimavi, D., Wolski, P., 2006. Long-term variations of annual flows of the Okavango and Zambezi Rivers. Phys. Chem. Earth 31 (15–16), 944–951.
27. Meier, P., Frömelt, A., Kinzelbach, W., 2011. Hydrological real-time modelling in the Zambezi river basin using satellite-based soil moisture and rainfall data. Hydrol. Earth Syst. Sci. 15, 999–1008.
28. Mertens, J., Liechti-Cohen, T., Matos, J.P., Tilmant, A., Schleiss, A.J., Wehrli, B., 2013. Adapted reservoir management in the Zambezi river basin to meet environmental needs. Int. J. Hydropower Dams 2, 80–84.
29. Milly, P.C.D., Betancourt, J., Falkenmark, M., Hirsch, R.M., Kundzewicz, Z.W., Lettenmaier, D.P., Stouffer, R.J., 2008. Stationarity is dead: whither water management? Science 319, 573–574.
30. Monteith, J.L., 1965. Evaporation and environment. Symp. Soc. Exp. Biol. 19, 205–234.
31. Mukosa, C., Pitchen, G., Cadou, C., 1995. Recent hydrologicaltrends in the Upper Zambezi and Kafue basins. In: Matiza, T., Crafter, S., Dale, P. (Eds.), Water Resources Use in the Zambezi Basin. Proceedings of a workshop held at Kasane. Botswana, April 28–May 2, 1993. IUCN, Gland, pp. 85–98.
32. Nash, J.E., Sutcliffe, J.V., 1970. River flow forecasting through conceptual models: Part 1 – a discussion of principles. J. Hydrol. 10, 282–290.
33. Nikulin, G., Jones, C., 2011. Simulated precipitation within the CORDEX-Africa framework: evaluation phase. In: EGU General Assembly 2011. Geophysical Research Abstracts (13): EGU2011-10315.
34. Paeth, H., Hall, N.M.J., Gaertner, M.A., Alonso, M.D., Moumouni, S., Polcher, J., Ruti, P.M., Fink, A.H., Gosset, M., Lebel, T., Gaye, A.T., Rowell, D.P., Moufouma-Okia, W., Jacob, D., Rockel, B., Giorgi, F., Rummukainen, M., 2011.
35. Progress in regional downscaling of west African precipitation. Atmos. Sci. Lett. (12), 75–82. Piani, C., Weedon, G.P., Best, M., Gomes, S.M., Viterbo, P., Hagemann, S., Haerter, J.O., 2010.
36. Statistical bias correction of global simulated daily precipitation and temperature for the application of hydrological models. J. Hydrol. 395, 199–215. Refsgaard, J.C., Henriksen, H.J., 2004. Modelling guidelines – terminology and guiding principles. Adv. Water Resour. 27, 71–82.

37. Scipal, K., Scheffler, C., Wagner, W., 2005. Soil moisture-runoff relation atthe catchment scale as observed with coarse resolution microwave remote sensing. Hydrol. Earth Syst. Sci. 9, 173–183.
38. Singh, R., van Wekhoven, K., Wagener, T., 2013. Hydrologic impacts of climate change in gauged and ungauged watersheds of the Olifants basin: a trading space-for-time approach. Hydrol. Sci. J. 59 (1), 29–55.
39. Stanzel, P., Nachtnebel, H.P., 2010. Mögliche Auswirkungen des Klimawandels auf den Wasserhaushalt und die Wasserkraftnutzung in Österreich. Oesterr. Wasser Abfallwirtsch. 9–10 (10), 180–187.
40. Tilmant, A., Beevers, L., Muyunda, B., 2010. Restoring a flow regime through the coordinated operation of a multireservoir system: The case of the Zambezi River basin. Water Resour. Res. 46, 11.
41. Tyson, P.D., Cooper, G.R.J., McCarthy, T.S., 2002. Millennial to multi-decadal variability in the climate of southern Africa. Int. J. Climatol. 22 (9), 1105–1117. Wilby, R.L., 2010. Evaluating climate model outputs for hydrological applications – opinion. Hydrol. Sci. J. 55 (7), 1090–1093.
42. Winsemius, H.C., Savenije, H.H.G., Gerrits, A.M.J., Zapreeva, E.A., Klees, R., 2006. Comparison of two model approaches in the Zambezi River basin with regard to model reliability and identifiability. Hydrol. Earth Syst. Sci. 10, 339–352.
43. Winsemius, H.C., Savenije, H.H.G., Bastiaanssen, W.G.M., 2008. Constraining model parameters on remotely sensed evaporation: justification for distribution in ungauged basins? Hydrol. Earth Syst. Sci. 12, 1403–1413.

CITATION

Harald Kling, Philipp Stanzel, Martin Preishuber, Impact modelling of water resources development and climate scenarios on Zambezi River discharge, Journal of Hydrology: Regional Studies, Volume 1, July 2014, Pages 17-43, ISSN 2214-5818, http://dx.doi.org/10.1016/j.ejrh.2014.05.002.

CHAPTER 3

Remote Sensing of the Ecology and Functioning of the Mekong River Basin with Special Reference to the Tonle Sap

Simon Nicholas Benger

Flinders University, Australia

INTRODUCTION

The management of large transnational river basins is subject to a range of challenges stemming from differing national priorities, governance of land use activities and resource use, and differences in institutional capacity, data gathering and data sharing. Over vast, often inaccessible areas, remote sensing allows for rapid assessment of ecological resources and hydrological processes. This includes quantification of the extent and ecological functioning of vegetation communities, defining the distribution, duration and timing of flooding, measurement of water quality parameters, groundwater assessment, habitat assessment, and predictive modelling of the ecological impacts of landuse activities and changes to hydrological cycles. Remote Sensing technologies currently allow unparalleled capability for environmental monitoring and management. Data recording and delivery systems, sensor platforms, and sensor technology are constantly improving and each year deliver better remote sensing products for a wide array of applications. Largely independent of geopolitical constraints and boundaries, remote sensing systems allow investigation and analysis of water resources and ecosystem functioning and processes at a range of scales. Large transnational river basins such as the Mekong River basin, can be studied in their entirety or in part.

This chapter examines the use of remote sensing techniques in various investigations in the Mekong River Basin, with particular reference to work on the Tonle Sap (Great Lake) of Cambodia.

The Mekong River Basin

The Mekong is the 10th largest river basin in the world in terms of mean annual outflow, with an annual discharge of 475 billion m^3 (Daming, 1997). From its source on the Tibetan Plateau, it flows some 4,800 km south to the Mekong Delta in Vietnam, draining a total catchment area of 795,000 km^2(MRC, 2005). The Mekong River Basin spans the six countries of China, Myanmar, Lao PDR, Thailand, Cambodia and Vietnam and forms the major hydrological resource for Southeast Asian. The basin has always faced the challenges of widespread poverty, increasing demands on water and environmental resources, and conflict throughout the region (Jacobs, 2002). There is lack of coordinated management of the basin, although the Mekong River Commission (MRC), and its predecessors the Mekong Committee and the Mekong Interim Committee have sought to foster dialogue between the member countries since the late 1950s. The main achievement of the MRC, however, has been the development in recent decades of an extensive data gathering and dissemination system, flood forecasting and warning systems, and advancing the understanding of the ecological and physical attributes of the basin (Jacobs, 2002).

Flow and runoff in the Mekong is strongly seasonal, reflecting the influence of the annual monsoon in the lower reaches of the basin. The wet season peaks in September-October with flows in the lower basin of 20,000-30,000 m^3s^{-1}, compared to dry season flows of approximately 2,000 m^3s^{-1}, which are derived mainly from snow melt in the upper basin (Mekong Secretariat, 1989). The Mekong is subject to natural annual variability which affects the size of the flood peak in any given year and is driven primarily by El Nino Southern Oscillation (ENSO) events (Kiem et al. 2004). Future flood pulse activity may be threatened, however, with significant water resources development occurring throughout the Mekong basin, along with the uncertain effects of climate change on precipitation and river flows. Development and water impoundment and extraction upstream on the Mekong, particularly in southern China but also in Laos, Thailand and Vietnam, is thought to be affecting the size, timing and intensity of the monsoonal flood pulse (Blake, 2001; Osbourne, 2006). Although catchments in China account for approximately one fifth of the flows in the Mekong overall, they can contribute 70-80% of flows during the dry season (MRC, 2005). The two main dams built by China on the upper reaches of the Mekong are the Manwan dam, which was completed in 1993, and the larger Dachaosan dam, which was completed in 2003. Campbell et al. (2006) show a reduction in average flood height and flooded area over the past decade. One of the most significant hydrological features of the lower reaches of the Mekong basin is the Tonle Sap lake in Cambodia, which fills annually and plays an important role in flood attenuation and sediment and nutrient exchange from the Mekong (MRC, 2005). Events occurring in the upper reaches of the Mekong that systematically alter the flood hydrograph or change its timing are likely to have significant effects on the sustainability of the Tonle Sap (Kummu et al. 2004).

The Tonle Sap

The Tonle Sap or Great Lake of Cambodia (Figure 1) forms part of a unique and ecologically significant sub-system within the Mekong basin. It is the largest freshwater lake in Southeast Asia, covering an area of 250,000-300,000 Ha during the dry season and up to 1.6 million Ha during the wet season (ADB, 2002). Expansion of the lake during the wet season is due primarily to the annual monsoonal flood pulse moving down the Mekong and entering the lake through the Tonle Sap River, which reverses its course as the water level in the Mekong rises above that of the lake. Besides drainage from the Mekong during the monsoonal flood, 13 other catchments drain into the lake. The lake plays an important role in flood peak attenuation and flow control to the Mekong Delta, storing up to 40 km^3 of Mekong floodwater each year and releasing it slowly back into the system (MRC, 2005). It was listed as a UNESCO Biosphere reserve in 1997, and is designated as a Protected Area under Cambodian Royal decree and through numerous international agreements. By far the largest area of savannah swamp forest and inundated forest in Asia, it contains important Ramsar-listed wetlands, and supports extensive fisheries and agriculture of critical importance to the Cambodian economy. Some 2.9 million people live in the five provinces surrounding the lake (ADB, 2002). With economic and political stability returning to the region in the past decade, the population around the margins of the lake is expanding rapidly, along with agricultural activity. Floodplain hydrology and wetland, flooded forest and riparian communities are being modified at a rapid rate and with major ecological impacts.

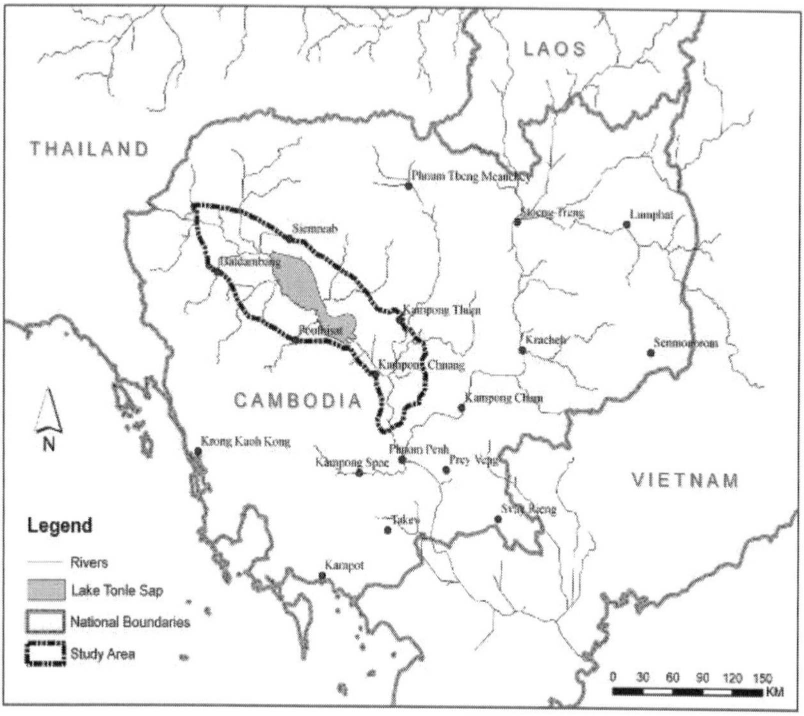

Figure 1. The Tonle Sap Floodplain Study Area.

Several ancient Ankorean capitals flourished on the northern margins of the Tonle Sap floodplain between A.D. 802 and 1431 (Chandler, 1996) and the floodplain has been modified in the past. Structures were built to control the movement of water across the floodplain and harvest it for agricultural purposes or to provide sites for aquaculture activities. This resulted in an extensive hydraulic network estimated to cover more than 1000 km^2 (Evans et al. 2007). Centralised control under the Ankorean court and highly organised agricultural production across the lowlands around Angkor produced economic surpluses for the state. Construction of reservoirs and channels occurred on a large scale through controlled use of labour, including slaves (Higham 2001). Although much of this original irrigation and agricultural infrastructure has probably been subsumed into more recent schemes, or obliterated by the flooding cycles of the lake, several examples of the largest ancient structures remain. These include the Domdek channel: a 200m wide channel extending approximately 80km through the floodplain with 10-15 m high walls; and the Western Baray at Angkor; a water storage covering 17.5 km^2.

The inflow from the Mekong accounts for approximately 70 % of flow into the Tonle Sap lake (Penny, 2006), with the remainder coming from local catchments. Some 80 % of the sediments and nutrients entering the lake from the Mekong are retained (MRC, 2005) and this annual process supports floodplain and fisheries productivity. The Tonle Sap lake is therefore highly susceptible to changes in the size, timing and duration of the annual monsoonal flood pulse, whether that occurs as a result of climate change or upstream water resources development. The past decade has seen reductions in flood height and flooded area of the lake (Campbell et al. 2006), although 2008 saw larger than normal floods throughout the Mekong. Kiem et al. (2008) in their latest modelling, suggest that precipitation will increase by 4.2% on average throughout the Mekong basin, concentrated in the upper sections of the basin in China. Chinvanno (2003) suggests that while there will be some shift in the timing of the flood peak, flooding durations will still be adequate for the survival of significant wetland areas on the Tonle Sap.

Most management efforts on the Tonle Sap to date have focussed on maintaining the lake's fisheries, which provide up to 70% of the protein intake for the entire Cambodian population (van Zalinge et al. 2000), and protection of the Ramsar wetlands as bird nesting sites. Natural resource management is severely under-resourced and occurs in a piecemeal manner (Bonhuer and Lane, 2002) in the face of poorly delineated jurisdictions and conflicting economic interests. Despite the importance of the Tonle Sap lake to the Cambodian economy, only in recent years have authorities and research agencies begun to characterise the flooding cycles of the lake or map floodplain vegetation distributions. Some modelling of lake hydrology was completed in 2003 (Koponen et al. 2003) and an Asian Development Bank project is currently underway to produce GIS datasets of lake resources (ABD, 2002). The Cambodian Mekong National River Commission (MNRC) in association with the multi-country Mekong River Commission (MRC) now monitor flood conditions in the Mekong and the Tonle Sap, but data is restricted to a limited number of gauging stations and is often not reliable. For example, the nearest MRC gauging station is located at Kampong Chhnang, on the Tonle Sap tributary (Figure 1).

REMOTE SENSING OF FLOODPLAIN STRUCTURES

Many extensive water impoundment structures as part of irrigation schemes have been built throughout the Tonle Sap floodplain to retain flood waters and support dry season rice cropping. Such anthropogenic modification of the floodplain occurs primarily on the northern margins of the lake in closer proximity to larger settlements. It is likely that these structures have a significant impact on floodwater distribution and movement and will simply serve as flood barriers if peak lake levels are diminished. Floodplain structures result in permanent inundation of large areas that were previously subject to a wetting and drying cycle; essential for the maintenance and survival of many plant and animal species, including many economically important fish species. In addition, retention and restriction of floodwater movement inhibits nutrient exchange between the floodplain and the lake, and movement of juvenile fish into the lake and the Mekong. The impoundments disrupt the moving littoral of the lake's flood pulse (Junk et al. 1989) where high turnover rates of organic matter and nutrients occur. The gradient of plant species adapted to seasonal degrees of inundation, nutrients and light no longer experiences the conditions under which it evolved.

An aim of the current study was to use remote sensing to determine the extent of floodplain structures around the Tonle Sap and where they lay in relation to flooding extent and duration. Major structures associated with irrigation schemes located within the annually flooded zone of the floodplain were mapped using WAAS corrected GPS to an accuracy of 2-3 m during fieldwork in 2006. High resolution Japanese/NASA ASTER (Advanced Spaceborne Thermal Emission and Reflection Radiometer) imagery was obtained over the floodplain for a range of wet and dry-season dates. ASTER senses in 14 spectral bands in the visible, shortwave and thermal infrared, at 15 m, 30 m and 90 m resolutions respectively (Lillesand et al. 2008). From the 37 ASTER multi-spectral surface reflectance product images obtained for the study, a mosaic of 11 dry-season images covering the Tonle Sap floodplain was constructed with rectification carried out using GCPs (Ground Control Points) collected during fieldwork. The available ASTER coverage over the Tonle Sap is fragmented, both spatially and temporally, due to almost perpetual high levels of cloud cover, but it was possible to generate a near-complete mosaic (Figure 2). As most of the structures occurred on the northern shore of the lake, generally they tended to have an east-west orientation. Horizontal spatial filtering was carried out on the imagery to identify and map the extent of major structures. Spatial filters operate on an image to emphasise or deemphasize image data of varying spatial frequencies. Directional first differencing is a simple directional image enhancement technique which improves the delineation of linear features (Lillesand et al. 2008).

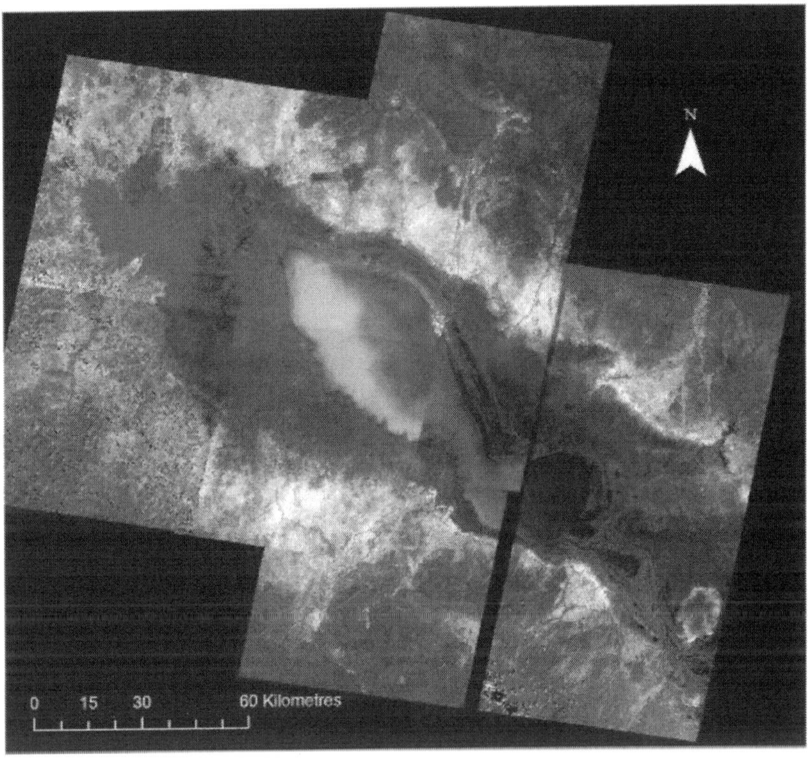

Figure 2. ASTER Dry Season Image Mosaic of the Tonle Sap Floodplain.

Using the filtered images it was possible to identify and map approximately 321 km of major impoundment structures which directly affect water movement across the floodplain (Figure 2). These were generally constructed parallel to the lake shoreline and serve to retain large volumes of water behind them as the lake waters recede after October in any given year. Major structures are defined as being greater than approximately 2m in height, although there are large networks of smaller formal and informal dykes, weirs and regulators which are also used or have been used to modify water movement. Most were built by hand during the Khmer Rouge years using forced labour, and in the absence of any hydrological or engineering knowledge (Kiernan, 1996). Extensive colonisation of these structures with floodplain vegetation has meant that they now form permanent features on the floodplain. According to the flood cycle patterns revealed by the time series analysis described later in this chapter, most of the impoundment structures are built within the zone that would normally be inundated around the end of August in any given year, drying out by mid-December, giving a flood residence time of around 3-4 months (Figure 3). There is also an obvious interaction with floodplain soils. Significant waterlogging occurs around these structures for much of the year, which is a commonly observed phenomenon associated with water storages (Ramireddygari et al. 2000). This is causing a number of changes to wetlands in these areas. Euphorbiaceae, Fabaceae, and Combretaceae species, which once colonised the mosaic of flooded savannah forest are being replaced by those which can tolerate saturated soils. In the areas behind the dyke walls, which now form permanent water storages, natural wetland species have disappeared completely, due either to blanket infestations of water hyacinth and fringing introduced scrub species.

A secondary impact can also be observed. Irrigated rice fields are present on the lake shore side of most water impoundment structures. Increased nutrient levels associated with the application of fertilisers to the rice fields are likely to be affecting the surrounding wetlands through mobilisation during flooding in the wet season and affecting groundwater quality. Leaching of nutrients into the groundwater from these areas, along with increased utilisation of the groundwater by wetland plants due to higher groundwater levels has created succession towards more nutrient tolerant weeds such as*Mimosa pigra* (Campbell et al. 2006). Similarly, pesticides leaching into groundwater which lies close to the surface are affecting the wetland soils which contain the eggs of hundreds of fish species deposited when the lake is in flood. Changes in predator prey relationships that are important for the ecology of the lake (Scheffer, 1998) and its fisheries are likely to be occurring due to floodwater containment. The impoundments would restrict movement of larger fish into shallow areas of lake for predation during flooding and also form barriers to movement of juveniles out from hatchery zones to the lake and the Mekong system. This undoubtedly contributes to the well-documented reduction in the number of fish species and changes in size of individuals (Puy et al. 1999; Bonheur & Lane, 2002). However, the impoundments are also an important source of protein for the occupants of the floodplain, as they effectively operate as large unmanaged aquaculture sites for much of the year, possibly reducing pressure on lake fish stocks.

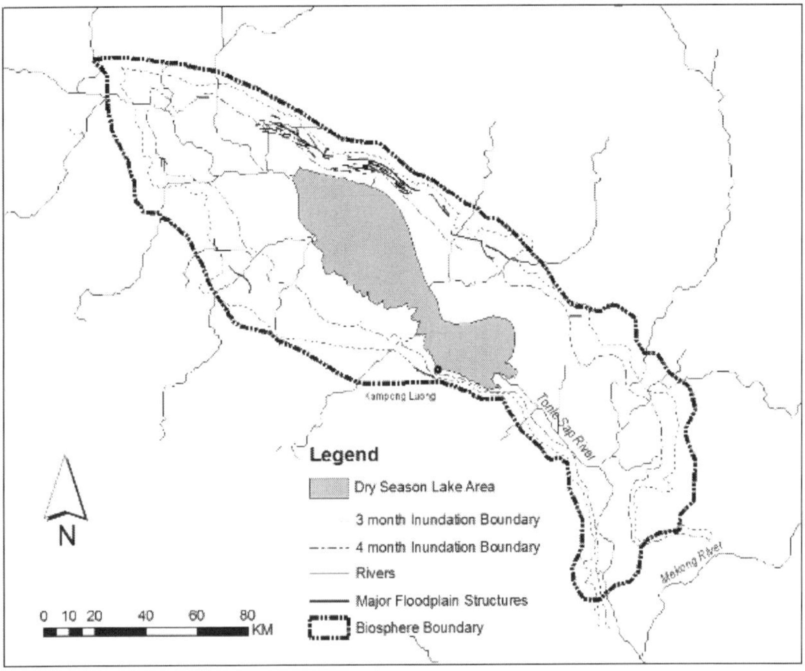

Figure 3. Major Water Impoundment and Barrier Structures on the Tonle Sap Floodplain.

The Khmer Rouge under Pol Pot sought to dramatically increase the areas of land under cultivation on the floodplain, and emptied the cities to provide forced labour for the extensive irrigation schemes that were established (Kiernan, 1996). These structures form by far the largest spatial extent of modifications to the present day floodplain, although many have now been abandoned or arc in partial usc. Of those surveyed during fieldwork, approximately 40% are now in disuse and others partially used on a seasonally varying basis depending on flooding extent, land availability and population pressures (Bonheur & Lane, 2002). Many of the areas originally modified for rice cultivation have failed to be maintained by the present population because of their inaccessible locations within the floodplain, poor siting, lack of centralised management and maintenance of the schemes, and destruction of infrastructure by flooding. Many of these areas have now reverted to permanent wetlands in areas that would previously have dried out when the floodwaters receded each year. Wright et al. (2004) report on some 570 irrigation schemes existing within the Tonle Sap basin, with only 195 being fully operational today. It is not know how many of these schemes fall within the area of the floodplain, although it is likely that a proportion are located in non-flooded areas. A recent phenomenon on the floodplain is the development of large scale privately owned irrigation schemes which seek to harvest floodwaters for rice production. Substantial areas of floodplain previously utilised by village communes for lower impact agricultural activities are being modified in this way. The use of ring-dyke structures to harvest flood waters for rice production are seen as the ideal new model for agricultural development of the Tonle Sap floodplain (Someth et al. 2009).

REMOTE SENSING OF GROUNDWATER RESOURCES

Remote sensing has been widely used to measure the moisture content of soils (Jensen, 2007), although this often depends on the soil grain size and mineralogy, which will affect the ability of a soil mass to store water. Recently, a number of studies have begun to examine the use of remote sensing for inferring the nature of groundwater resources. Brunner et al. (2007) provide an overview of the potential use of remote sensing in the provision of data to support groundwater modelling in a number of large river basins. Other examples of recent studies include Mutiti et al. (2008), who examined groundwater resource development potential using Landsat imagery, Hendricks Franssen et al. (2008) who inferred groundwater patterning from remotely sensed data, and Milzow et al. (2009) who examined groundwater and hydrology of the large river/wetland system of the Okavango Delta using remote sensing. A range of remote sensing technologies are available to assist in the study of groundwater resources. These include technologies such as radar, LIDAR and digital photogrammetry to derive elevation products, airborne EM (electromagnetics) to examine changes in electrical conductivity in the shallow subsurface, and the remote sensing of vegetation, salt crusts and other surface features as a proxy for subsurface groundwater conditions (Brunner et al. 2007).

Groundwater resources are particularly important for the region in and around the Tonle Sap floodplain, as they form the major water supply for human use (Wright et al. 2004). The sedimentary depression of the Tonle Sap is surrounded by low-lying alluvium, with older coarser ferruginous silts, sands and grits around the perimeter overlain by red-clayey and silty sediments (Stanger et al. 2005). The alluvial deposits of the Tonle Sap floodplain are believed to be very good shallow aquifers, with high recharge rates (5-20 m^3/h) and a groundwater table generally within 4-6m of the surface. Groundwater quality is generally good apart from high iron content reducing palatability in some areas, and dangerous levels of arsenic contamination in others (Wright et al. 2004). In response to the large amplitude floods that characterise the hydrological cycle of the Tonle Sap, there is an annual cycle in groundwater levels from depths of around 6 m in riparian areas to a few centimetres in some parts of the floodplain (Stanger et al. 2005).

Loss of vegetation, particularly deep rooted tree species, reduces uptake of water from the soil profile and exacerbates waterlogging problems in the wetlands. A large seasonal population usually migrates from upland areas and the non-flooded areas of the Tonle Sap basin to the floodplain as the floodwaters recede, building temporary settlements on and around the water impoundment structures (Bonheur & Lane, 2002). The temporary settlements facilitate activities such as dry season rice cropping and fishing and informal aquaculture. Human settlement compounds the loss of larger wetland tree species in these areas as they form the primary source of fuelwood and building materials. This occurs on a wide scale despite a complete ban on all forms of timber extraction from the flooded forest areas. As well as the loss of deep rooted tree species, groundwater levels are also likely to be affected by the permanent and semi-permanent water impoundments, which would have a subsurface connection to the local water table (Ramireddygari et al. 2000). An aim of the current study was to investigate whether these effects existed and were detectable using available optical remotely sensed imagery. Soil moisture absorbs incident radiant energy in the 1.4, 1.9 and 2.7 μm regions, although the spectral response can be complex depending on soil type and soil characteristics (Jensen, 2007).

In three fieldsite locations on the Tonle sap floodplain, the relationship between groundwater and water storages was examined. During fieldwork elevated soil waterlogging adjacent to water storages could be observed through the presence of dark saturated soils along with consequent changes in vegetation type. Trenches were dug adjacent to the structures to ascertain depth to water table, and these confirmed water tables lying at or near the surface. Remotely sensed analysis of Landsat imagery over these areas made it possible to map the extent of waterlogging extending out from these structures. This involved the generation of wetness index maps, using the Kauth-Thomas (KT) transformation (Kauth & Thomas, 1976; Collins & Woodcock, 1996). A wetness index map derived from the

KT transformation will indicate not only the level of surface soil moisture, but also the wetness of associated vegetation (Mutiti et al. 2008). A wetness index map for the fieldsite locations examined is presented in Figure 3. While the results do indicate a relationship between the size of the water storage and the area detected, such results are difficult to interpret without further information on the quantity of water stored, the duration of storage, the soil types and localised topography, all of which are unavailable for the Tonle Sap Floodplain. However, they did indicate the potential of remote sensing to detect and quantify these effects, and demonstrate the effects of waterlogging of soils adjacent to water impoundment structures – an important consideration given the rapid agricultural development occurring in some areas of the floodplain utilising water impoundments.

REMOTE SENSING OF FLOODPLAIN VEGETATION

The monsoonal driven flood pulse fills the lake and floods an extensive area of the floodplain, usually for several months from August through to January, creating a unique flooded forest plant community (McDonald et al. 1997). These temporary wetlands serve essential ecosystem processes in terms of nutrient exchange between the lake (and the Mekong system upstream) and the floodplain, and are essential for fish breeding (Puy et al. 1999). Flooded forests are found mainly around the dry-season lake shoreline and comprise about 10% of the floodplain and are dominated by *Barringtonia acutangula*, *Baringtonia micratha* and *Diospyros cambodiana.* At higher elevations are extensive areas of short tree shrubland dominated by species of Euphorbiaceae, Fabaceae, and Combretaceae, together with *Barringtonia acutangula* (Wikramanayake & Dinerstein 2001) and seasonally flooded sedgelands and grasslands occupy the distal margins. Large seasonal contrasts in lake levels affect the characteristics of the wetland vegetation (Penny, 2006), with some forest areas enduring fluctuations of up to 8m and complete canopy submergence for months at a time (McDonald et al. 1997).

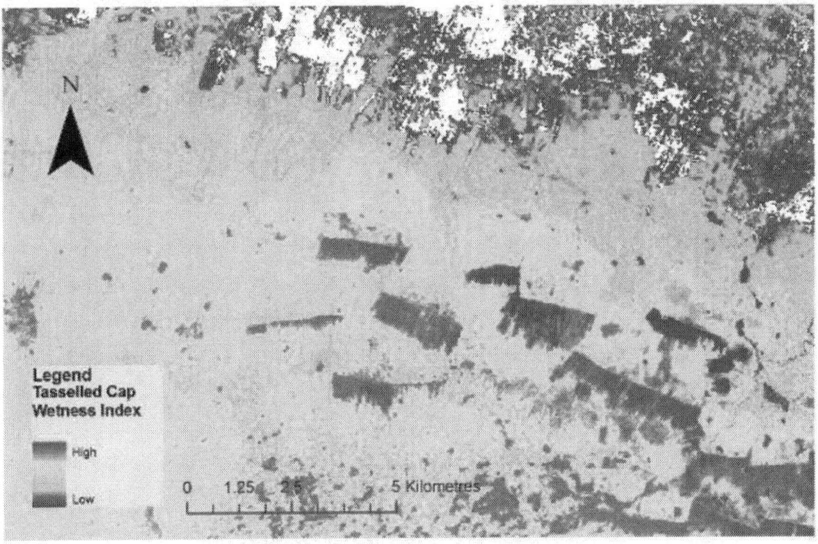

Figure 4. KT Wetness Index Map of an area of the Tonle Sap Floodplain.

Like many ephemeral wetlands around the world, the distribution of the mosaic of flooded forest, scrub and grassland around the lake is determined largely by the duration and depth of flooding (Bonheur & Lane, 2002), and to a lesser degree substrate. The tropical climate, nutrient rich soils and abundant water present on the floodplain mean that vegetative growth occurs rapidly, and forests and wetlands quickly regenerate. It has been suggested that much of the present wetland vegetation is secondary regrowth (McDonald et al. 1997), although this seems unlikely over the majority of the Tonle Sap wetlands, which are highly inaccessible. Only on the northwestern margins of the lake, where the ancient civilizations of Angkor flourished between A.D. 802 and 1431 (Chandler, 1996) is large scale clearing likely to have occurred to facilitate extensive agricultural schemes. Many of these were re-established, often unsuccessfully, during the Khmer Rouge period (Kiernan, 1996). While limited historical data exists on the distribution of plant communities across the floodplain, the majority of the floodplain vegetation is still intact, although often modified in areas closer to settlements, and capable of near normal ecological functioning subject to floodwater availability.

One aim of the current study was to use remotely sensed data to map the current distribution of wetland and floodplain vegetation around the Tonle Sap, so that these could be examined in relation to where they occur spatially on the floodplain in relation to flooding extent and duration.

Previous efforts to map wetland vegetation distribution and its relationships to flooding (e.g. Kite 2001) utilised generalised USGS landcover classifications not particularly suited to the Tonle Sap floodplain. The ADB has generated landuse maps and metrics for all of the Tonle Sap catchments (ADB, 2009), although the landuse categories used are generalised and non-species specific.

Image classification procedures can be used to identify, map and quantify vegetation units of interest in remotely sensed imagery. The overall objective of image classification procedures is to automatically categorise all pixels in an image into land cover classes or themes (Lillesand et al. 2008). Image classification attempts to use the spectral information available in the data for each pixel as the numerical basis for categorisation. Different feature types will manifest different combinations of spectral response in each band (depending on sensor type) based on their inherent spectral reflectance and emittance properties which may also be variant in space and time. Spectral pattern recognition refers to the family of classification procedures that utilise this pixel-by-pixel spectral information as the basis for automated land classification (Lillesand et al. 2008).

Supervised classification is the procedure most often used for quantitative analysis of remote sensing image data. It rests upon using suitable algorithms to label the pixels in an image as representing particular ground cover types or classes (Richards & Jia, 2006). The multidimensional normal distribution of a spectral class is specified completely by its mean vector and its covariance matrix. Consequently, if the mean vectors and the covariance matrices are known for each spectral class then it is possible to compute the set of probabilities that describe the relative likelihoods of a pattern at a particular location belonging to each of those classes (Lillesand et al. 2008). It can then be considered as belonging to the class which indicates the highest probability. Therefore, if the mean vectors and the covariance matrix are known for every spectral class in an image, every pixel in the image can be examined and labelled corresponding to the most likely class on the basis of the probabilities computed for the particular location for a pixel. Before that classification can be performed however, the mean vectors and covariance matrix are estimated for each class from a representative set of pixels, called a training set. These are pixels which the analyst knows as coming from a particular spectral class.

Supervised classification consists therefore of three broad steps. First a set of training pixels is selected for each spectral class using the reference data available in the form of digital vegetation maps. The second step is to determine the mean vectors and covariance matrices for each class from the training data. This completes the learning phase. The third step is the classification stage, in which the relative likelihoods for each pixel in the image are computed and the pixel labelled according to the highest likelihood (Richards & Jia, 2006). Numerous mathematical approaches have been developed for spectral pattern recognition and it is beyond the scope and relevance of this chapter to review them all. Some commonly used classifiers are the Minimum-Distance-to-Means, parallelepiped, Gaussian Maximum Likelihood Classifier (MLC) and the Piecewise Linear Classifier. In the current study, MLC was used for the supervised classification of the ASTER optical imagery. MLC has a demonstrated reliability in achieving accurate classification of land cover types across a range of different environments (Bolstad & Lillesand, 1991; San Miguel-Ayanz & Biging, 1997).

MLC is one of the most commonly used supervised classification methods and it has been demonstrated to be extremely powerful and efficient in a great number of investigations (Maselli et al. 1990). It works most effectively when dealing with normal distribution in the spectral data, although it has also been shown to be relatively resistant to class distribution anomalies (Hixson et al. 1980; Yool et al. 1986). This classifier quantitatively evaluates both the variance and the covariance of the category spectral response patterns. It assumes a Gaussian distribution in the category training data, which is generally a reasonable assumption. Using this assumption the distribution of a category response pattern can be completely described by the mean vector and covariance matrix, it is possible to compute the statistical probability of an unknown pixel belonging to particular land cover class. In essence the maximum likelihood classifier delineates ellipsoidal "equiprobability contours" in the scatter diagram of spectral values which act as the decision regions (Lillesand et al. 2008).

The main limitation of maximum likelihood classification is the large number of computations required to classify each pixel. This is particularly true when either a large number of spectral channels are involved or a large number of spectral classes must be differentiated. Numerous extensions and refinements of the maximum likelihood classifier have been developed (Lillesand et al. 2008). These include the use of lookup tables in which the category identity of all possible combinations of digital numbers is determined prior to classifying the image and each unknown pixel is classified simply by reference to these lookup tables. This avoids the need to carry out complex statistical calculations for each pixel as they have already been determined for each category. Another means of optimizing maximum likelihood classifiers is to use some method to reduce the dimensionality of the dataset used to perform the classification. Procedures such as the principal components, canonical components (Jensen and Waltz, 1979) and tassled cap (Kauth and Thomas, 1976) transformations achieve this reduction of the dataset by making use only of the significant sections of the data.

Floodplain vegetation type and distribution were observed and mapped in the field in and around the Tonle Sap through a number of fieldwork surveys conducted in 2005 and 2006, in accessible locations. Remote sensing offers the ability to map landcover types over large areas, based on spectral information collected from representative vegetation communities and other landuse and landcovers (Lillesand et al. 2008). On the basis of training sites mapped throughout the floodplain during fieldwork, wetland vegetation and landcover across the floodplain was classified into 20 classes using maximum likelihood classification on the 9 visible/near-infrared and shortwave infrared bands of the ASTER imagery, which were resampled to 30 m. This facilitated determination of the types and extent of wetland vegetation directly affected by the water impoundment structures and their relationship to flooding patterns. Classification accuracy was assessed using standard confusion matrices to generate overall accuracy and Kappa statistics (Congalton & Green, 2008), using one training site for each landcover type not used in the original classification. As a result of the classification carried out over the floodplain using the imagery, it was possible to generate floodplain metrics for the various vegetation and landuse classes, and these are presented in Table 1.

Table 1. Landcover classification results for the Tonle Sap floodplain.

Vegetation/Landuse Class	Area (ha)	Percentage
Barringtonia acutangula dom.	107928	7.75
Barringtonia acutangula dom.	765737	55.01
Diospyros cambodiana dom.	184344	13.24
Euphorbiaceae Shrubland	53794	3.86
Tiliaceae Shrubland	61062	4.39
Mimosa pigra	7551	0.54
Sedge	17810	1.28
Phragmites Reeds	36928	2.65
Thornbush	3380	0.24
Water storage, unvegetated	19257	1.38
Water storage, vegetated	11345	0.81
Agricultural - rice	34972	2.51
Agricultural - fallow	11839	0.85
Legume cropping	250	0.02
Grasslands	67410	4.84
Mudbanks saturated soil	2722	0.2
Bare dry soil	2344	0.17
Firescar	1569	0.11
Rock outcrop	183	0.01
Human settlement	1664	0.12
Total (excluding)	1392089	100
Kappa = 0.83		

RELATIONSHIPS BETWEEN ELEVATION AND FLOODPLAIN VEGETATION

High quality digital elevation data are essential for the assessment of floodplains and spatial arrangement of vegetation communities. Numerous studies of wetland vegetation have suggested that elevation is a primary determinant of vegetation type and location within wetland systems (Scoones, 1981; Hughes, 1990), and substrate to a lesser degree. Analysis of elevation data can yield important information on the spatial arrangement of vegetation in wetland and floodplain environments as it determines the extent and duration of flooding of these areas. In many of the developing countries which comprise the Mekong River basin high quality survey data is simply not available, and over large inaccessible areas such as the Tonle Sap floodplain, ground based survey is logistically impossible. Therefore remote sensing offers the primary means of gathering such data.

There are a range of remote sensing techniques available for the generation of elevation data or digital elevation models (dems). In general, higher precision in these products is accompanied by higher cost of acquisition and processing. Techniques include digital stereo photogrammetry, radar interfereometry and LIDAR. For the Mekong basin the primary dataset that has been utilised is the United States Geological Survey (USGS) GTOPO 30 dem, which is a 30 arc-second resolution product. The Shuttle Radar Topography Mission (SRTM) global product can also be used which has a 3 arc-second (approximately 90 m) resolution with 5 m vertical accuracy (Slater et al. 2006), and more recently, the ASTER GDEM global dem became available in 2009 with 30 m resolution and 15 m vertical accuracy. Of these datasets, only the latter is suitable for use in a low relief environment such as the Tonle Sap floodplain. In the GTOPO 30 and SRTM data, variations in floodplain relief are dominated by data anomalies. In all cases where remotely sensed elevation data are available, finer resolution dem data can be interpolated, but these may lead to a false representation of precision as they will normally retain the errors present in the original data (Longley et al. 2007). Kite (2001) used the USGS GTOPO 30 product for hydrological modelling of the Mekong Basin, and the ADB (2009) show flooded area maps and metrics for the Tonle Sap catchments interpolated from contour maps. In the current study, remotely sensed elevation data was utilised to investigate the relationship between the primary wetland and floodplain vegetation types and elevation, and hence relationship to flooding. For this purpose the ASTER GDEM product was used after processing to remove anomalies, most of which occur over areas of open water and along tile edges, and extracting only elevations below 30m in height. The resultant 30m dem for the Tonle Sap floodplain is shown in 3D in Fig 5.

Figure 5. 3D dem of Tonle Sap floodplain, derived from ASTER GDEM.

A simple GIS-based analysis of the location of vegetation communities in relation to elevation yields information on the elevation ranges they occupy within the floodplain. While elevation alone is not the sole determinant of flooding effects on vegetation, in a floodplain such as the Tonle Sap, where overbank flooding from the lake is the primary source of floodwater, it does indicate the sensitivities of various ecological communities to water level ranges. The depth and duration of flooding for these communities is a primary determinant of their evolution in a given location and their ecological functioning (Campbell et al. 2006). The main vegetation classes and their elevation ranges derived from this analysis are shown in Table 2.

Table 2. Elevation ranges for primary vegetation classes.

Vcgctation Class	Min imum E levation (m)	Max imum Elevation (m)	(m)
Barringtonia acutangula dom.	0.6	1.8	1.2
Barringtonia acutangula dom.	1.9	8.2	6.3
Diospyros cambodiana dom.	2.4	8.7	6.3
Euphorbiaceae Shrubland	5.3	10.4	5.1
Tiliaceae Shrubland	6.2	12.6	6.4
Sedge	8.2	1 1 . 3	3.1
Phragmites Reeds	1	2 .6	1.6
Grassland	9.1	18.4	9.3

The elevation ranges of the primary vegetation classes of the Tonle Sap confirm that the flooded forest and reed communities occupy lower elevations on the floodplain, with Savannah woodland communities at higher elevations followed by shrubland, sedge and grasslands. Flooded forest, reed and sedge communities occupy the narrowest elevation ranges on the floodplain, while those communities at higher elevations are most likely to be affected by reductions in flood height. This information can then be used with the information on temporal flood extent patterns described below to characterise the horizontal and vertical arrangement of species on the floodplain. This landscape ecology approach to the understanding of floodplain structure provides important information on ecological functioning. Landscape ecology is based on the hypothesis that the interactions among biotic and abiotic components of the landscape are spatially mediated. Not only are the flows of energy material or species from place to place affected by the locations of the places in the landscape, but these flows then determine the interactions among energy, material and species (Malanson, 1993). A central theme of landscape ecology is that spatial structure controls the processes that continuously reproduce the structure. Landscape ecology is an approach to the study of the environment that emphasizes complex spatial relations. The relative locations of phenomena, their overall arrangement in a mosaic and the types of boundaries between them, become the priorities of study (Forman & Godran, 1986; Ingegnoli, 2002).

FLOOD DETECTION AND MAPPING

The monsoonal flood pulse is the primary mechanism affecting productivity in the Tonle Sap lake, wetlands and floodplain. The economically important fisheries of the Tonle Sap are strongly influenced by the maximum flooded area and resultant area of fish feeding and breeding habitat (Webby et al. 2005). Remote sensing of the inundation patterns across the study area therefore formed an important part of the current study. Knowledge of the extent and residence time of floodwaters on the floodplains of major rivers is essential for hydrological and biological studies of these systems, and yet for most areas of the Mekong, this remains largely unknown beyond simple maps of flood extent. For the Tonle Sap, the ADB has compiled maps showing minimum and maximum flood extents for the catchments around the lake as derived from satellite image interpretation (ADB, 2009). For the areas examined in this study, information on flow rates and stream heights may be available, but because of the low relief and complex hydrology of many wetland areas, these data do not correlate well with inundation patterns. Rates of organic matter production, decomposition and export to the river channel are closely linked to floodplain inundation patterns. Primary production rates in inland wetlands are very high and these communities may cover hundreds of thousands of square kilometres (Matthews and Fung, 1987). In many large river systems with associated extensive wetland areas, the difficulty in determining the extent of flooding makes it difficult to accurately estimate wetland area and characterise vegetation relationships. Ground measurement of flooding in forested wetlands is severely limited by the inaccessibility typical of these areas, where mobility is often hampered by flooding and boggy conditions. Remote sensing offers the ability to detect flooded over such areas, and this is typically done using optical or radar imagery.

With regard to optical remote sensing of inundation and the spectral reflectance of water, probably the most distinctive characteristic is the absorption of energy at near-infrared (NIR) wavelengths. Locating and delineating water bodies with remote sensing data is done most easily at NIR wavelengths because of this property (Lillesand et al. 2008). However, various conditions of water bodies manifest themselves primarily in visible wavelengths. Landsat TM imagery has been used to map floodwater distribution and characteristics (e.g Imhoff et al. 1987; Pope et al. 1992; Mertes et al. 1993, 1995;Johnston and Barson 1993), and optical SPOT data has also been used for floodwater mapping (Blasco et al. 1992).

Remote sensing of flooding may also be hampered by forest canopies that render the land/water boundary invisible to infrared and visible wavelength sensors and by frequent cloud cover during periods of rainfall. These limitations are largely overcome by SAR radar sensors which are unaffected by clouds and can significantly penetrate relatively dense forest canopies (Hess et al. 1990). Passive microwave remote sensing has also proved useful for revealing large-scale inundation patterns, even in the presence of cloud cover and dense vegetation (Choudery 1991, Sippel et al. 1994). The bright appearance of flooded forests on radar images results from double-bounce reflections between smooth water surfaces and tree trunks or branches. Enhanced back scattering at L-band has been shown to occur in a wide variety of forest types and is a function of both stand density and branching structure (Hess et al. 1990). Steep incidence angles (20-30°) are optimal for detection of flooding, since some forests exhibit bright returns only at steeper angles. Backscattering from flooded forests is enhanced by underlying water. For forests of moderate density, L-band returns are dominated by corner reflections between trunks and surface and between branches and surface (Richards et al. 1987a). Scattering from a smooth water surface is specular, whereas that from soil includes a significant diffuse component and therefore the amplitude of returns will be higher for standing water beneath forests.

There is a high degree of structural diversity associated with flooded forests, as they occur on numerous substrates, in both saline and fresh water and at a wide range of latitudes (Matthews and Fung 1987). Most frequently studied have been the swamp forests of the coastal plains of the southeastern United States. Relatively bright L-band returns from semi-permanently to permanently flooded stands have been reported in several studies (e.g. Hoffer et al. 1986, Evans et al. 1986, Wu and Sader 1987). Detection of underlying water in mangrove swamps was demonstrated by Imhoff et al. (1987) in the Sundarbans region of Bangladesh and by Ford and Casey (1988) in East Kalimantan. Bright returns for seasonally inundated temperate forests are described by Richards et al. (1987b) for Eucalyptus camaldulensis forests in Australia. Ford et al. (1986) distinguished flooded varzea forest from non-flooded forest using SIR-B scenes of the Rio Japura in the Amazon Basin.

The forest stands cited above have very diverse structures: canopy depth relative to total tree height, dominant branching angle, and crown shape are quite variable. They also encompass a wide range of leaf type and tree heights. It is clear that stands with low stem densities may appear bright at L-band (Hess et al. 1990). Enhancement has also been shown for stands described as dense or thick (Hoffer et al. 1986, Ford and Casey 1988). Enhanced backscattering from flooded forests thus occurs over a broad range of tree species, canopy structures and stand densities. Richards et al. (1987b) demonstrated that brighter returns from flooded forests are not simply a function of vegetation differences between upland and lowland sites. They were able to clearly distinguish between flooded and non-flooded portions of a single forest type.

The accuracy of flood detection using radar imagery is difficult to determine since most studies of flooded forests focus only on those areas which do yield bright L-band returns. Near or complete absence of backscatter from flooded Maryland swamps with dense canopies has been noted by Krohn et al. (1983). It appears that dense undergrowth may significantly affect double-bounce returns. Ford and Casey (1988) found the opposite to be true, however, in flooded mangrove forests of Kalimantan. They found that open stands of low slender trees did not yield bright returns on SIR-B imagery while adjacent denser mangrove stands did. The above examples suggest that for certain forest types, the extent of flooding beneath the canopy would be underestimated using L-band radar. Overestimation would occur if other targets yielding bright returns were mistaken for flooded forests. Other sources of bright returns would normally be able to be visually distinguished from flooded forest based on shape, pattern, associated features and minimal site knowledge. A more serious source of confusion is non-forest vegetation naturally occurring adjacent to flooded forest. Flooded marshes (emergent herbaceous vegetation) typically appear dark at L-band (Krohn et al. 1983, Ormsby et al. 1985). However, marsh vegetation sometimes yields bright returns very similar to those from flooded forests (Krohn et al. 1983).

The magnitude of enhancement associated with double bounce beneath flooded forests can vary significantly. In many studies, variations in magnitude appear to be the result of differences in stand composition as well as flooding (Hess et al. 1990). The problem of separating backscatter variation caused by differences in vegetation from that caused by flooding was minimised in the study byRichards et al. (1987b), because of the virtually monospecifc stands of eucalyptus examined. Backscattering from flooded and non-flooded sites within the forest was estimated to vary by 10.8 dB: a substantial difference. Treating the canopy as a uniform layer of small particles, Engheta and Elachi (1982) estimate the enhancement resulting from the presence of a perfectly reflecting surface beneath the canopy to be 3 to 6 dB. It appears from the literature that L-band radar imagery used in the current study should enable accurate delineation of floodwater boundaries.

Aims of this study in relation to flood detection and mapping were to utilise remote sensing methods to (a.) characterise the flood cycles of the lake; (b.) map the spatial distribution of water across the floodplain, and; (c.) determine the relationship between the flooding cycles of the lake and vegetation distributions across the floodplain. The current flood monitoring and mapping efforts of the MNRC and MRC rely on simple linear models of the relationship between river gauge height collected at only a few locations and maximum annual volume and flooded area. Few of the tributaries which drain the 13 catchments around the lake and make significant contributions to lake volume and flooded area have any gauging stations, and hydrological relationships between these tributaries and the lake are complex (Penny, 2006).

For the current study, regional scale MODIS (Moderate resolution Imaging Spectrometer) data was used to determine inundation patterns. MODIS images in 36 spectral bands at 250 m, 500 m and 1 km resolutions, dependent on wavelength, and is widely used for multiple land and ocean applications which require high frequency temporal coverage (Lillesand et al. 2008). A large time-series of MODIS 500m 8 Day Surface Reflectance imagery collected over the period 2001-2005 was used to characterize the flood cycles during the period June to March, at weekly intervals, where the data was of sufficient quality. The MODIS imagery was subsetted to the area of the Tonle Sap and rectified to the ASTER basemap (Figure 2) with its much higher spatial precision using 6 GCPs per image.

The temporal dynamics of the flooded area for the lake are affected by landcover, infiltration rates, and local catchment inputs and cannot be estimated simply from lake gauge height. Inundation mapping in floodplain environments can be problematic due to the presence of high levels of vegetative cover, shallow inundation over large areas and dark organic rich alluvial soils which can appear inundated when they are not (Pearce, 1995). The methods used to map inundation can have a marked effect on the observed patterns (Frazier et al. 2003). On the Tonle Sap, the use of AIRSAR and JERS-1 radar data has been investigated as means of mapping inundation at localised scales (Milne & Tapley, 2005) but this has not been applied at the scale of the entire floodplain. Usual inundation mapping methods using optical imagery involve use of a ratio of mid-infrared reflectance to a visible band reflectance (Lillesand et al. 2008) although this is generally only suitable for relatively deep water. Investigations of techniques for floodplains suggest a combination approach using this ratio and mid-infrared (MIR) change detection is necessary to deal effectively with the shallow water problem (Sims, 2004). Due to the unique nature of the floodplain vegetation and shallow inundation over much of the wet season lake area, a specialised flood detection algorithm was developed for the Tonle Sap using MODIS B6/B4 ratio combined with a B1 threshold, and the accuracy of the technique was verified using the wet-season ASTER imagery. An example of the output from this analysis for a single image date is shown in Figure 6. The MODIS time series was used to determine the extent of flooding and flood duration in conjunction with hydrological data from the CNMC and Mekong River Commission.

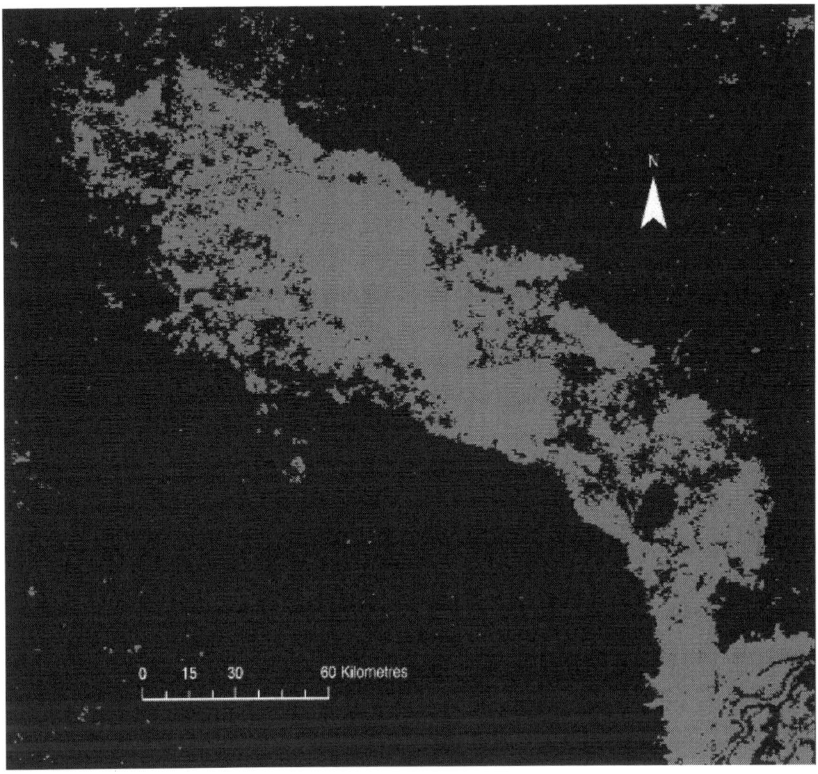

Figure 6. MODIS derived flood extents for the Tonle Sap – an example.

The MODIS derived flood maps indicate a reduction in flooding extent of the Tonle Sap lake since 2000. While the 2000 flood was large by historical standards, and caused widespread damage and loss of life throughout the Mekong Basin (CNMC, 2006), every year since then has been characterised by a reduction in the spatial extent of flooding across the floodplain, apart from the 2008 flood for which the MODIS imagery products are not yet available. This corresponds with MNRC and MRC observations that the flood peaks are now reduced in amplitude and have a much faster fill and drain cycle (CNMC, 2006). Some authors have suggested large dam development throughout the Mekong, and particularly in China, may be responsible (Blake, 2001). The very large Dachaosan dam in southern China began filling in 2003. The monsoons deliver large quantities of water very quickly into the dams where it can be released slowly throughout the year for hydroelectricity generation and for irrigation. The Chinese government currently has another three dams under construction in the upper reaches of the Mekong, with the Xiaowan dam now nearing completion, and another three are at the planning stage (Osbourne, 2006). This will form an 8 dam cascading system capable of retaining very large volumes of water that would otherwise contribute to the monsoonal Mekong flood pulse. With limited fossil fuel reserves and exponential growth in energy demand, the Mekong and other Chinese rivers are seen as offering abundant cheap and clean power. The Chinese dams in the upper reaches of the Mekong are unlikely to be responsible for all reduced flow into the Tonle Sap, as the region may also be experiencing some ongoing effects of drought and climate change (MRC, 2005), and irrigation development is also occurring rapidly on other tributaries which feed the lake. Other current and proposed dams for Laos, Thailand and Vietnam are likely to further ameliorate the Mekong flood pulse in the future. The output from the analysis of the MODIS time-series was then used to model the effects of inundation variability on the wetland and floodplain vegetation on the Tonle Sap floodplain.

GIS MODELLING OF THE EFFECTS OF FLOODING CHANGES ON VEGETATION

The MODIS time-series for the period 2001-2005 shows the area of the Tonle Sap flooded each year and duration of inundation. A goal of the current study was to be able use all the remotely sensed data and derived information on the functioning and spatial arrangement of vegetation and landuse on the floodplain to predict what changes might occur due to interference with the annual flood pulse. This entailed determining the flooding characteristics of floodplain vegetation in terms of depth, timing and duration of flooding and relating these to the spatial distribution of changes in flood patterns. The effects of possible diminished flood peak height and duration on the floodplain were simulated by using an average dry year hydrograph averaged from the four years 1992, 1993, 1999 and 2003 from CNMC data for the Tonle Sap to modify the maximum flood extent model derived from the MODIS imagery. Increasing water use and extraction throughout the Mekong is likely to create move toward dry year conditions with reduced water availability. The average dry year extent was subtracted from the average maximum flood extent derived from the five years of MODIS data for the period 2001-2005, and processed at the resolution of the floodplain dem (30 m). The results show the likely changes in the extent of flooding on the Tonle Sap floodplain if the flooding was likely to be reduced to drier year conditions due to water resource development in the Mekong Basin (Figure 7). When used with the vegetation and landuse cover classifications of the floodplain, this enables GIS modelling of the changes likely to occur in respective landcover types due to reductions in flooded area.

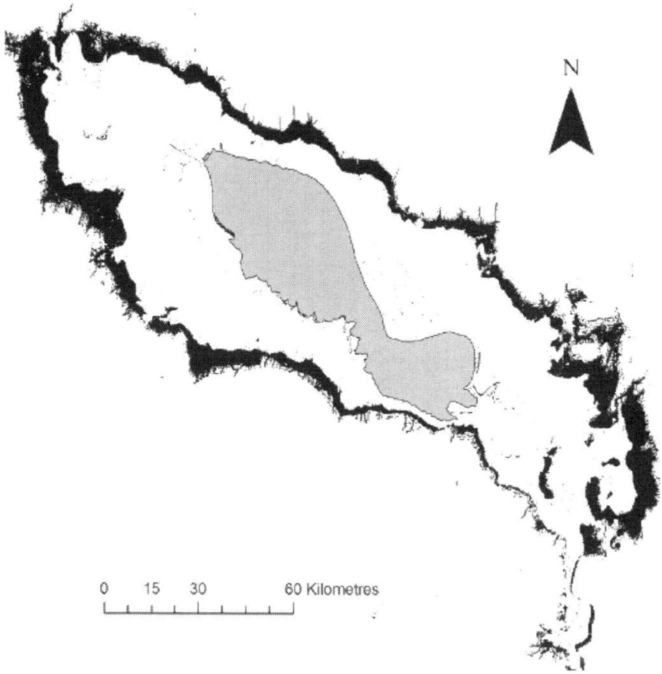

Figure 7. Modelled reduction in flooded area from MODIS derived flood extents, with change shown in black.

The temporal dynamics of the annual flood event on the Tonle Sap are also revealed in the MODIS data, and show the variability in the duration of the flooding. A cell-based GIS analysis was used to calculate the change in the duration of flooding between the simulated dry year conditions and average conditions. This approach also allows for the determination of how changes occur over temporal cycles, making it possible to develop a dynamic means of estimating changes to vegetation and landuse types. The temporal flood cycle model derived from the MODIS time-series and the dem data were integrated into ArcGIS 9.3 ModelBuilder (Environmental Systems Research Institute, 2009). The time series data provides weekly time steps showing change in flooded area. The mean duration of inundation per cell of landcover type was then generated and this data is summarised in Table 3, showing the change in flood residence time for primary vegetation units between the simulated dry season flood and an average flood. Results indicate that the largest reductions in flood duration will be experienced by the Savannah woodland communities, followed by the shrubland communities, with minor change in the sedge and grassland communities. The core wetland areas of flooded forest and reeds occur at lower elevations and show no reduced flood duration in this analysis. The results of simulating reduced flooding based on the average dry year hydrological data for the Tonle Sap from the GIS-based flood extent model indicate that reductions in flood peak and duration such as those experienced during dry years will have a significant effect on inundation area of the floodplain. This would result in reduction in flooded area reducing from 13,286 km^2 to 11,134 km^2, or approximately 16%.

Table 3. Change in flood durations between simulated dry year and average flood conditions for primary vegetation classes

Vegetation Class	Simulated Dry Year Flood Duration (days)	Average Flood Duration (days)	Change (days)
Barringtonia acutangula dom.	318	318	0
Barringtonia acutangula dom.	242	311	69
Diospyros cambodiana dom.	213	288	75
Euphorbiaceae Shrubland	58	79	21
Tiliaceae Shrubland	51	68	17
Sedge	43	52	9
Phragmites Reeds	324	324	0
Grassland	11	23	12

CONCLUSION

Remote sensing is able to provide valuable information on the structure, processes and functioning of the Tonle Sap floodplain. Large, inaccessible wetland and floodplain systems such as the Tonle Sap can be studied from space with a range of remote sensing technologies in combination with appropriate fieldwork and reference data. Interference with the natural flood cycles and inundation patterns of the lake and surrounding floodplain are causing changes in vegetation and are likely to be affecting the biological productivity not only on the Tonle Sap but throughout the Mekong system. The myriad impacts occurring in and around the impoundment structures on the floodplain are changing wetland community composition and structure, which in turn will affect fisheries productivity and species biodiversity. Local livelihoods are already affected by fierce (often violent) competition for lake and floodplain resources (Bonheur & Lane, 2002), and as the wetlands and floodplain degrade further this is likely to increase. Historical development of water resources has had significant impacts on the environments and catchments in parts of the floodplain, and caused permanent changes in the hydrology of these areas (Kummu, 2009), and this will continue and accelerate with population growth in the region. Water resource use upstream of the Tonle Sap is potentially reducing and moderating the monsoonal flood pulse which sustains the lake and floodplain system. This may be linked to the timing of large dam construction within the Mekong River basin, although Laos and Thailand are extracting increasing amounts of water from the Mekong as well for use in rapidly expanding rice irrigation schemes (Osbourne, 2006). While social benefits may arise from amelioration of floods which in some years can cause extensive property damage and loss of life, this must be balanced against the need to maintain flood cycles which can sustain the environment of the Tonle Sap, and economic activities such as fishing and agriculture. There is an urgent need to develop effective cross-border management plans and agreements for the water resources of the Mekong system before the unique and economically important Tonle Sap region slips into further decline.

Future events in the Mekong basin, whether related to climate change or human development, will have important ramifications for the Tonle Sap. The annual flood pulse which sustains lake and floodplain ecology is vulnerable to change and as it changes the primary vegetation communities on the Tonle Sap floodplain will most likely face significant declines. In addition, in-situ impacts from upstream developments in the sub-catchments of the lake, as well as further modification of the floodplain will act to reduce water availability and wetland area. The floodplain is already exhibiting signs of over-exploitation (Campbell et al. 2006) and this will increase in line with population and development pressures. It is critical that future basin planning and water resource extraction between the Mekong Basin countries be coordinated in order to preserve the size, duration and timing of the flooding of the Tonle Sap.

REFERENCES

1. Asian Development Bank 2002 Report and Recommendation for the Tonle Sap Environmental Management Project, ADB Report RRP: Cam 33418.
2. Asian Development Bank 2005 Summary Initial Environmental Examination Report for the Tonle Sap Sustainable Livelihoods Project in Cambodia, August 2005, ABD.

3. Asian Development Bank 2009 The Tonle Sap Initiative: Future Solutions Now, Available online:http://www.adb.org/Projects/Tonle_Sap/default.asp
4. S. N. Benger, 2006 Groundwater interactions with the wetlands of the Tonle Sap, Cambodia, in Proc. HydroEco 2006, Karlovy Vary, Czech Republic, Sept 2006, 45 48 . 8-09036-351-2
5. D. Blake, 2001 Proposed Mekong Dam scheme in China threatens millions in downstream countries. World Rivers Review 4-5, 43 51, 08906211
6. F. Blasco, M. F. Bellan, M. U. Chaudhury, 1992 Estimating the extent of floods in Bangladesh using SPOT data, Remote Sensing of Environment 39, 167 178 , 0034-4257
7. N. Bonheur, B. D. Lane, 2002 Natural resources management for human security in Cambodia's Tonle Sap Biosphere Reserve, Environmental Science and Policy 51, 33 42 , 1462-9011
8. P. V. Bolstad, T. M. Lillesand, 1991 Rapid maximum likelihood classification, Photogramm. Eng. Remote Sens. 57, 67 74, 0034-4257
9. P. Brunner, H. J. Hendricks Franssen, L. Kgotlhang, P. Bauer-Gottwein, W. Kinzelbach, 2007 How can remote sensing contribute to groundwater modelling? Hydrogeology Journal 15(1), 5 18, 1431-2174
10. Cambodia National Mekong Committee (CNMC) 2006 Cambodia Country Report: Flood information in Cambodia, Proceedings of the 4th Annual Mekong Flood Forum "Improving Flood Forecasting and Warning Systems for Flood Management and Mitigation in the Lower Mekong Basin", Siem Reap, Cambodia, May 2006, 23 36 .
11. I. C. Campbell, C. Poole, W. Giesen, J. Valbo-Jorgensen, 2006 Species diversity and ecology of Tonle Sap Great Lake, Cambodia, Aquatic Sciences-Research Across Boundaries 68, 355 373, 1015-1621
12. D. Chandler, 1996 A History of Cambodia, Westview Press Inc, 9-74710-065-7
13. B. J. Choudery, 1991 Passive microwave remote sensing contribution to hydrological variables, Surveys in Geophyics 12, 63 84 , 0169-3298
14. J. B. Collins, C. E. Woodcock, 1996 An assessment of several linear change detection techniques for mapping forest mortality using multitemporal Landsat TM data. Remote Sensing of Environment 56 66 77 , 0034-4257
15. R. G. Congalton, K. Green, 2008 Assessing the Accuracy of Remotely Sensed Data: Principles and Practices, 2nd Edition, CRC Press, 978-1-42005-512-2 Boca Raton, FL.
16. H. Daming, 1997 Facilitating regional sustainable development through integrated multi-objective utilization management of water resources in the Lancang-Mekong river basin, Journal of Chinese Geography 7, 4, 1861-9568

17. N. Engheta, C. Elachi, 1982 Radar scattering from a diffuse vegetation layer over a smooth surface, IEEE Trans. Geosci. and Remote Sens. 20, 212 216 , 0196-2892
18. Environmental Systems Research Institute 2009 ArcGIS 9.3 ModelBuilder Software, ESRI, Redlands, CA.
19. D. Evans, C. Pottier, R. Fletcher, S. Hensley, I. Tapley, A. Milne, M. Barbetti, 2007 A comprehensive archaeological map of the world's largest preindustrial settlement complex at Angkor, Cambodia, Proceedings of the National Academy of Sciences of the United States of America 104 (36), 14277 14282 , 1091-6490
20. D. E. Evans, T. G. Farr, J. P. Ford, T. W. Thompson, C. L. Werner, 1986 Multipolarisation radar images for geological mapping and vegetation discrimination, IEEE Trans. Geosci. and Remote Sens. 24, 246 257 , 0196-2892
21. J. P. Ford, D. J. Casey, 1988 Shuttle radar mapping with diverse incidence angles in the rainforest of Borneo, Int. Journal of Remote Sensing 5, 927 943 , 1366-5901
22. J. P. Ford, J. B. Cimano, B. Holt, M. R. Ruzek, 1986 Shuttle Imaging Radar Views the Earth from Challenger: The SIR-B Experiment, Jet Propulsion Laboratory publication 86-10, Pasadena, California
23. R. T. T. Forman, M. Godron, 1986 Landscape Ecology, Wiley, 0-47187-037-4 New York.
24. P. Frazier, K. Page, J. Louis, S. Briggs, A. I. Robertson, 2003 Relating wetland inundation to river flow using Landsat TM data. Int. Journal of Remote Sensing 2419, 3755 3770 , 1366-5901
25. H. J. Hendricks Franssen, P. Brunner, P. Makobo, W. Kinzelbach, 2008 Equally likely inverse solutions to a groundwater flow problem including pattern information from remote sensing images, Water Resources Research, 44, W01419, doi:10.1029/2007WR006097
26. L. L. Hess, J. M. Melack, D. S. Simonett, 1990 Radar detection of flooding beneath the forest canopy: a review, Int. Journal of Remote Sensing 11, 1313 1325 , 1366-5901
27. C. Higham, 2001 The Civilisation of Angkor, Orion Books, London. 1-84212-584-2
28. K. Hixson, D. Scholz, N. Funs, 1980 Evaluation of several schemes for classification of remotely sensed data, Photogramm. Eng. Remote Sens. 46, 1547 1553 , 0099-1112
29. R. M. Hoffer, D. F. Lozano-Garcia, D. D. Gillespie, P. W. Mueller, M. J. Ruzek, 1986 Analysis of multiple incidence angle SIR-B data for determining forest stand characteristics, The Second Spaceborne Imaging Radar Symposium, JPL Publication 86-26, Pasadena CA, 159 164

30. F. M. R. Hughes, 1990 The influence of flooding regimes on forest distribution and composition in the Tana River Floodplain, Kenya, Journal of Applied Ecology 27, 475 491 , 1365-2664
31. V. Ingegnoli, 2002 Landscape Ecology: A Widening Foundation, Springer, 978-3-54042-743-8 Amsterdam.
32. M. Imhoff, C. Vermillion, M. H. Story, A. M. Choudery, A. Gafoor, 1987 Monsoon flood boundary delineation and assessment using spaceborne imaging radar and Landsat data, Photogramm. Eng. Remote Sens. 53, 405 413 , 0099-1112
33. J. W. Jacobs, 2002 The Mekong River Commission: transboundary water resources planning and regional security, The Geographical Journal 168, 354 364 , 1861-9568
34. J. R. Jensen, 2007 Remote Sensing of the Environment, 2nd Edition, Pearson Prentice Hall, 0-13188-950-8 Upper Saddle River, NJ.
35. S. K. Jensen, F. A. Waltz, 1979 Principal Components Analysis and Canonical Analysis in Remote Sensing, in Proc. American Photogrammetric Society 45th Annual Meeting, 337 348 , 0099-1112
36. R. M. Johnson, M. M. Barson, 1993 Remote sensing of Australian wetlands: an evaluation of Landsat TM data for inventory and classification, Aust. J. Mar. Freshwater Res. 44, 235 252 , 0067-1940
37. W. J. Junk, P. B. Bayley, R. E. Sparks, 1989 The Flood Pulse Concept in river-floodplain systems, Can. Spec. Publ. Fish. Aquat. Sci. 106, 110 127 . 1205-7533
38. R. J. Kauth, G. S. Thomas, 1976 The Tasseled Cap- A graphic description of the spectral temporal development of agricultural crops as seen by Landsat, in Proc. LARS 1976 Symposium on Machine Processing of Remotely Sensed Data, Purdue University.
39. A. S. Kiem, H. Ishidaira, H. P. Hapuarachchi, M. C. Zhou, Y. Hirabayashi, K. Takeuchi, 2008 Future hydroclimatology of the Mekong River basin simulated using the high-resolution Japan Meteorological Agency (JMA) AGCM, Hydrological Processes 22 1382 1394 , 1099-1085
40. A. S. Kiem, H. P. Hapuarachchi, K. Takeuchi, 2004 Impacts of climate variability on streamflow in the Mekong River: an interesting challenge for hydrological modelling, In: Proc. River Symposium 2004, Sept 2004, Brisbane.
41. B. Kiernan, 1996 The Pol Pot Regime- Race, Power and Genocide in Cambodia under the Khmer Rouge, 1975-79, Yale University Press, 9-74710-043-6 New Haven
42. G. Kite, 2001 Modelling the Mekong: hydrological simulation for environmental impact studies, Journal of Hydrology 253, 1 13 , 0022-1694

43. J. Koponen, J. Josza, H. Lauri, J. Sarkkula, V. Markku, 2003 Modelling Tonle Sap Watershed and Lake Processes for Environmental Change Assessment, Mekong River Commission MRCS/WUP-FIN Model Report.
44. M. D. Krohn, N. M. Milton, D. B. Segal, 1983 Seasat synthetic aperture radar (SAR) response to lowland vegetation types in eastern Maryland and Virginia, J. Geophys. Res. 88, 1937 1952 , 0148-0227
45. M. Kummu, 2009 Water management in Angkor: Human impacts on hydrology and sediment transportation, Journal of Environmental Management 90, 3, 1413 1421, 0301-4797
46. T. M. Lillesand, R. W. Kiefer, J. W. Chipman, 2008 Principles of Remote Sensing and Image Analysis, 6th Edition, Wiley, 978-0-47005-245-7 New York.
47. P. A. Longley, M. F. Goodchild, D. J. Maguire, D. W. Rhind, 2005 Geographic Information Systems and Science, 2nd Edition, Wiley, 0-47087-000-1 New York.
48. G. P. Malanson, 1993 Riparian Landscapes, Cambridge University Press, 9780521384315, Cambridge
49. F. Maselli, C. Conese, G. Zipoli, M. A. Pittau, 1990 Use of error probabilities to improve area estimates based on maximum likelihood classifications, Remote Sensing of Environment 31, 155 160 , 0034-4257
50. J. McDonald, P. Bunnat, P. Virak, 1997 Plant Communities of the Tonle Sap Floodplain, UNESCO/IUCN/WI, Phnom Penh.
51. Mekong River Commission 2007 Annual Mekong Flood Report 2006, Mekong River Commission, 1728-3248 Vientiane.
52. Mekong River Commission 2005 Overview of the Hydrology of the Mekong Basin, Mekong River Commission, 1728-3248 Vientiane, November 2005.
53. Mekong Secretariat 1994 Annual Report 1994, Mekong Secretariat, 1728-3248 Bangkok.
54. L. A. K. Mertes, D. L. Daniel, J. M. Melack, B. Nelson, L. A. Martinelli, B. R. Forsberg, 1995 Spatial patterns of hydrology, geomorphology, and vegetation on the floodplain of the Amazon River in Brazil from a remote sensing perspective, Geomorphology 13, 215 232 , 0169-555X
55. L. A. K. Mertes, M. O. Smith, J. B. Adams, 1993 Estimating suspended sediment concentrations in surface waters of the Amazon River wetlands from Landsat images, Remote Sensing of Environment 43, 281 301, 0034-4257
56. A. K. Milne, I. J. Tapley, 2005 Change Detection Analysis in the Wetlands Using JERS-1 Radar Data: Tone Sap Great Lake, Cambodia, IEEE doi 0-7803-9119-5/05. 146 150

57. C. Milzow, N. Kgotlhang, P. Bauer-Gottwein, P. Meier, W. Kinzelbach, 2009 Regional review: the hydrology of the Okavango Delta, Botswana- processes, data and modelling, Hydrogeology Journal, 1431-2174 Published Online DOI 10.1007/s10040-009-0436-0
58. S. Mutiti, J. Levy, C. Mututi, N. S. Guturu, 2008 Assessing Ground Water Development Potential Using Landsat Imagery, Groundwater, Published Online DOI 10.1111/j.1745-6584.2008.00524.x
59. J. P. Ormsby, B. J. Blanchard, A. J. Blanchard, 1985 Detection of lowland flooding using active microwave systems, Photogramm. Eng. Remote Sens. 51, 317 328 , 0099-1112
60. M. Osbourne, 2006 River at risk: The Mekong and the water politics of China and Southeast Asia, Lowy Institute for International Policy, 1-92100-402-9 New York.
61. B. Pearce, 1995 The compilation of regional flood maps using remote sensing techniques over the Ballonne river catchment and downstream areas. Technical Report. Queensland Department of Primary Industries, Brisbane, QLD.
62. D. Penny, 2006 The Holocene history and development of the Tonle Sap, Cambodia. Quaternary Science Reviews 25, 310 322 , 0277-3791
63. K. O. Pope, E. J. Sheffner, K. J. Linthicum, C. L. Bailey, T. M. Logan, E. S. Kasischke, K. Birney, A. R. Nlogu, C. R. Roberts, 1992 Identification of the central Kenyan Rift Valley fever virus vector habitats with Landsat TM and evaluation of their flooding status with airborne imaging radar, Remote Sensing of Environment 40, 185 196 , 0034-4257
64. L. Puy, S. Lek, S. T. Touch, S-O. Mao, B. Chhouk, 1999 Diversity and spatial distribution of freshwater fish in Great Lake and Tonle Sap river Cambodia, Southeast Asia, Aquatic Living Resources 126, 379 386 , 0990-7440
65. S. R. Ramireddygari, M. A. Sophocleous, J. K. Koelliker, S. P. Perkins, R. S. Govindaraju, 2000 Development and application of a comprehensive simulation model to evaluate impacts of watershed structures and irrigation water use on streamflow and groundwater: the case of Wet Walnut Creek Watershed, Kansas, USA. Journal of Hydrology 2363-4, 223 246 , 0022-1694
66. J. A. Richards, X. Jia, 2006 Remote Sensing Digital Image Analysis- An Introduction, 4th Edition, Springer-Verlag, Berlin. 978-3-54025-128-6
67. J. A. Richards, G-Q. Sun, D. S. Simonett, 1987a L-band radar backscatter modelling of forest stands, IEEE Trans. Geosc. Remote Sens. 25, 487 498 , 0196-2892
68. J. A. Richards, P. W. Woodgate, A. K. Skidmore, 1987b An explanation of enhanced radar backscattering from flooded forests, Int. Journal of Remote Sensing 8, 1093 1100 , 1366-5901

69. J. San Miguel Ayanz, G. S. Biging, 1997 Comparison of single-stage and multi-stage classification approaches for cover type mapping with TM and SPOT data, Remote Sensing of Environment 59, 92 104 , 0034-4257
70. M. Scheffer, 1998 The Ecology of Shallow Lakes, Chapman and Hill, 0-41274-920-3, London
71. I. Scoones, 1991 Wetlands in Drylands: key resources for agricultural and pastoral production in Africa, Ambio 20, 366 371 , 0044-7447
72. N. Sims, 2004 The Landscape-scale Structure and Functioning of Floodplains, Unpublished PhD Thesis, University of Canberra.
73. S. J. Sippel, S. K. Hamilton, J. M. Melack, B. J. Choudery, 1994 Determination of inundation area in the Amazon River floodplain using the SMMR 37 GHz polarisation difference, Remote Sensing of Environment 48, 70 76 , 0034-4257
74. J. A. Slater, G. Garvey, C. Johnston, J. Haase, B. Heady, G. Kroenung, J. Little, 2006 The SRTM data "finishing" process and products. Photogramm. Eng. Remote Sens. 72(3), 237 247 , 0099-1112
75. P. Someth, N. Kubo, H. Tanji, S. Lyd, 2009 Ring dike system to harness floodwater from the Mekong River for paddy rice cultivation in the Tonle Sap Lake floodplain in Cambodia, Agricultural Water Management 96, 100 110 , 0378-3774
76. G. Stanger, T. Van Truong, K. S. Ngoc, T. V. Luyen, T. T. Thanh, 2005 Arsenic in groundwaters of the Lower Mekong, Environmental Geochemistry and Health 27, 341 357 , 1573-2983
77. N. Top, N. Mizoue, S. Kai, T. Nokao, 2004 Variation in woodfuel consumption patterns in response to forest availability in Kampong Thom Province, Cambodia, Biomass and Energy 27, 57 68 , 0167-5494
78. N. Van Zalinge, N. Thouk, T. C. Tana, D. Leung, 2000 Where there is water, there is fish? Cambodian fisheries issues in a Mekong River Basin perspective. In: Ahmed, M. and Hirsh, P. (Eds) Common Property in the Mekong: Issues of Sustainability and Subsistence. ICLARM Study Review.
79. R. Webby, P. T. Adamson, J. Boland, P. G. Howlett, A. V. Metcalfe, J. Piantadosi, 2005 The Mekong- Applications of Value at Risk (VaR) and Conditional Value at Risk (CVaR) simulation to the benefits, costs and consequences of water resources development in a large river basin. In: MODSIM 2005 International Congress on Modelling and Simulation. (ed. by A. Zerger & R.M. Argent), 2109 2115, 0-97584-000-2, Modelling and Simulation Society of Australia and New Zealand, December 2005, Brisbane.
80. E. Wikramanayake, E. Dinerstein, 2001 Terrestrial Ecoregions of the Indo-Pacific, Island Press, 1-55963-923-7 Washington DC.

81. G. Wright, D. Moffatt, J. Wager, 2004 Establishment of the Tonle Sap Basin Management Organisation: Tonle Sap Basin Profile, Cambodia National Mekong Committee, Asian Development Bank Report TA2412-CAM.
82. S. T. Wu, S. A. Sadler, 1987 Multipolaristaion SAR data for surface feature delineation and forest vegetation characterisation, IEEE Trans. Geosc. Remote Sens. 25, 67 76 , 0196-2892
83. S. R. Yool, Y. L. Star, J. E. Estes, E. B. Botkin, D. W. Eckardt, F. W. Davis, 1986 Performance analysis of image processing algorithms for classification of natural vegetation in the mountains of southern California, Int. Journal of Remote Sensing 7, 683 702 , 1366-5901

CITATION

Simon Nicholas Benger (2009). Remote Sensing of the Ecology and Functioning of the Mekong River Basin with Special Reference to the Tonle Sap, Geoscience and Remote Sensing, Pei-Gee Peter Ho (Ed.), ISBN: 978-953-307-003-2, InTech, DOI: 10.5772/8296.

CHAPTER 4

Wastewater Reuse for Irrigation — Practices, Safe Reuse and Perspectives

I. Navarro, A. Chavez, J.A. Barrios, C. Maya, E. Becerril, S. Lucario and B. Jimenez

[1]Institute of Engineering, UNAM, Mexico

INTRODUCTION

Wastewater for Irrigation and Its Role in Sanitation and Human Health

Human impact on water bodies has become relevant since water withdrawal, especially for agriculture, has resulted in overexploitation of rivers, lakes and aquifers. In most countries agriculture represents by far the largest use of water and worldwide, it represents about 70% of total withdrawal and 90% of water consumption [1]. To address this problem wastewater reuse has proven to be an alternative to reduce anthropogenic impacts [2]. In addition, raw wastewater reuse in agriculture is a valuable tool available to developing countries to control pollution and tackle the challenge of increasing food production in water scarce areas. The benefits of reusing water in agriculture are many and beyond doubt: it saves considerable amounts of first-use water that may be assigned to critical uses; it provides nutrients that may substitute chemical fertilizers, increasing soil fertility and crop yield, and reducing production costs; it makes it possible to expand agricultural land in arid areas; it is a relatively cheap disposal method for raw wastewater; and it may avoid pollution of surface water. In addition, it has been demonstrated that the health risks of reusing water in agriculture are minimal as long as its biological quality meets established criteria [3, 4].

However, it should be considered that there are potential negative impacts that may arise, such as soil salinization as well as soil and groundwater pollution with metals and organic compounds. In addition, the use of raw wastewater for crop production poses health risks due to its microbial content, especially bacteria, viruses and parasites, which produce a wide range of diseases since many of them may survive on the environment for long periods of time.

The main issue associated to wastewater reuse is related to public health and infection risks, either real (produces a disease) or potential (transmits the infection but the disease does not develop). Infection rates may be high, low or minimal, depending on the type of pathogen, the infective dose, and the susceptibility of the affected person (host). According to epidemiological studies over the last 20 years, when untreated wastewater is applied to land for crop production, there exist real infection risks caused by pathogens. To reduce such risks, control actions must be implemented, such as treating wastewater to comply with regulation limits; developing fast, cheap, easy, and efficient detection techniques; breaking the disease-infection cycle with medical treatment; and developing education campaigns for the population [4-6].

INTERNATIONAL AND REGIONAL GUIDELINES AND COUNTRY REGULATIONS FOR TREATED WASTEWATER REUSE IN IRRIGATION

Globally, wastewater reuse has become significant and this has encouraged many countries to develop local regulations to control water quality for reuse with the aim of reducing health and environmental risks. Due to different geographic, economic, and social characteristics, development of such regulations has been gradual and dissimilar among countries. Developed countries have worked on these regulations for several years and among them, the United States of America, and in particular the state of California, applied the first regulations on agricultural reuse in 1918 [7]. As a result, California has one of the most strict and complete regulatory frameworks. In addition, the United States Environmental Protection Agency (US EPA) has developed regulations and criteria for water reuse, which are used as a reference by many countries. Moreover, the creation of international organizations has led to the publication of general recommendations. The World Health Organization (WHO) published a series of four volumes that include information about agricultural irrigation and reuse in aquaculture [8]. At the same time, the Food and Agriculture Organization of the United Nations (FAO) has its own guidelines that are similar to those of the WHO [9]. WHO proposes limits for indicator bacteria (total or faecal coliforms) and helminth eggs when wastewater is reused for agriculture, considering that helminth eggs are highly resistant to treatment process and common in the environment.

In Europe, Mediterranean countries have detailed legislation about this topic, while countries with high water availability, such as Germany or the United Kingdom, do not regulate reuse as it is seldom practiced. Spain sets limits on faecal coliforms and nematodes based on their 1985 Water Law. In the Americas, countries like Brazil, Costa Rica, Chile, and Mexico have made progress in developing regulations focused mainly on restricting water reuse based on microbial content, including indicator bacteria, helminth eggs and some metals, allowing organic matter and nutrients that are beneficial for agriculture to be used on land.

In Africa, despite the fact that several countries face a water crisis, many of them have lax or non-existent regulations and thus wastewater reuse is practiced uncontrolled. However, countries like Tunisia have detailed guidelines that include physicochemical and biological parameters, as well as heavy metals, an approach shared by some Mediterranean countries like Saudi Arabia, Israel, and Jordan [9].

In contrast, countries like Palestine, Libya, and Afghanistan, which have low water availability, have not been able to develop their own standards due to political conflicts and generally use either FAO or WHO guidelines. Finally, Indian regulation is outdated (1974) and limits just a few parameters, even though it has the world's second largest population.

Usually, most regulations for reuse in irrigation establish limits on one or more of the following parameters: indicator bacteria (total or faecal coliforms), helminth eggs (intestinal worms), nematodes (a subgroup of helminthes), organic matter (as biochemical oxygen demand, BOD), dissolved and suspended solids, and heavy metals.

SUCCESSFUL CASES OF WASTEWATER REUSE FOR IRRIGATION

Demographic growth and economic development of emerging American countries has promoted the implementation of several agricultural reuse projects, some of which are summarized below.

In Mendoza, Argentina, an area known as Campo Espejo used to be irrigated with raw wastewater (2,000 ha) but currently, 129,600 m^3/d originating from stabilization ponds are supplied for the irrigation of 1900 ha [10].

Chile has several successful reuse projects such as the Maipo and Maipocho regions where 130,000 ha are irrigated [11]; Antofagasta where about 20,000 m^3/d of treated water are produced and 65 ha are irrigated; and Santiago de Chile with 110,000 ha that use reclaimed water mixed with first-use water [10].

In Mexico, the central and northern part of the country, where 80% of the population live, is considered arid or semi-arid, and cities like Ciudad Juarez irrigate 26,000 ha with approximately 400,000 m^3/d of reclaimed water. Additionally, the Mezquital Valley, covering more than 90,000 ha, is one of the largest areas in the world where agricultural reuse is practiced [12] and where a large wastewater treatment plant (35 m^3/s) is under construction to improve water quality for irrigation.

In South America, Peru encompasses a number of different biomes that range from Amazon rainforest (more than 50% of the country) to the west coast, an area with low precipitation where most of the population lives, and where small irrigation projects like San Agustin (535 ha) and Tacna (738 ha), have been developed [10].

CASE STUDY: TULA VALLEY, MEXICO

The Tula Valley (also known as Mezquital Valley) is one of the largest areas irrigated with untreated wastewater in the world. The fact that it has been receiving wastewater since the late XIX century, and specifically for irrigation since the early XX century, makes this a unique site with regard to wastewater reuse. A large number of studies have described several processes and phenomena that emerge out of this practice, including incidental aquifer recharge, increase in crop yield, and treatment of the wastewater by the soil (in the same way as a soil-aquifer treatment system, SAT).

Concurrently, other studies have found an increase in health risks after using untreated wastewater in agriculture due to the presence of pathogens, heavy metals, and organic compounds in soil and even in groundwater. This section summarizes the history and current situation of the Tula Valley to share some lessons learned for more than 100 years of water reuse.

Metropolitan Area of the Valley of Mexico and Tula Valley

The Metropolitan Area of the Valley of Mexico (MAVM) has a population of 21.2 million and covers the Federal District (Mexico City) and 60 surrounding Municipalities (59 located in the State of Mexico and 1 corresponding to the State of Hidalgo). Water use is estimated to be 82 m^3/s which includes 91% first use water and 9% reclaimed water. The rate of wastewater generation (including collected precipitation) ranges from 52 to 300 m^3/s depending on the season, and it is conveyed to the Tula Valley. The reason for this practice is that Mexico City is located in an endorheic (closed) basin from which wastewater and excess precipitation needs to be transported to avoid flooding. On average, 60 m^3/s are sent by gravity or pumping to the Tula Valley via four artificial exits (Figure 1): the Tajo de Nochistongo (deep cutting through Nochistongo Hill; 1607-1789); the Tequixquiac Tunnel (1900); the New Tequixquiac Tunnel (1955) and the Central Emitter (1975). Currently, a fifth exit (East Emitter) with a capacity of 150 m^3/s is being constructed along 62 km and will be operating by 2016.

The Tula Valley is located 100 km north of Mexico City, with an elevation that ranges from 1,700 mamsl in the north, to 2,100 mamsl in the south. The climate is semiarid with an annual precipitation of 550 mm (national average is 790 mm) mainly between May and October [10] which contrasts with evapotranspiration that reaches 1,524 mm/y.

Originally, soils were low in organic matter and nutrients, and with such low precipitation, productivity was also low. By the end of the XIX Century, the Tula Valley started receiving raw wastewater from the MAVM. The first documented use of wastewater for irrigation in areas closed to the Salado River was reported around 1896. However, officially, reuse initiated in 1889 when wastewater was used to produce energy at the

Juandhó and La Cañada hydroelectric plants [13], and in 1912 for irrigation [14].

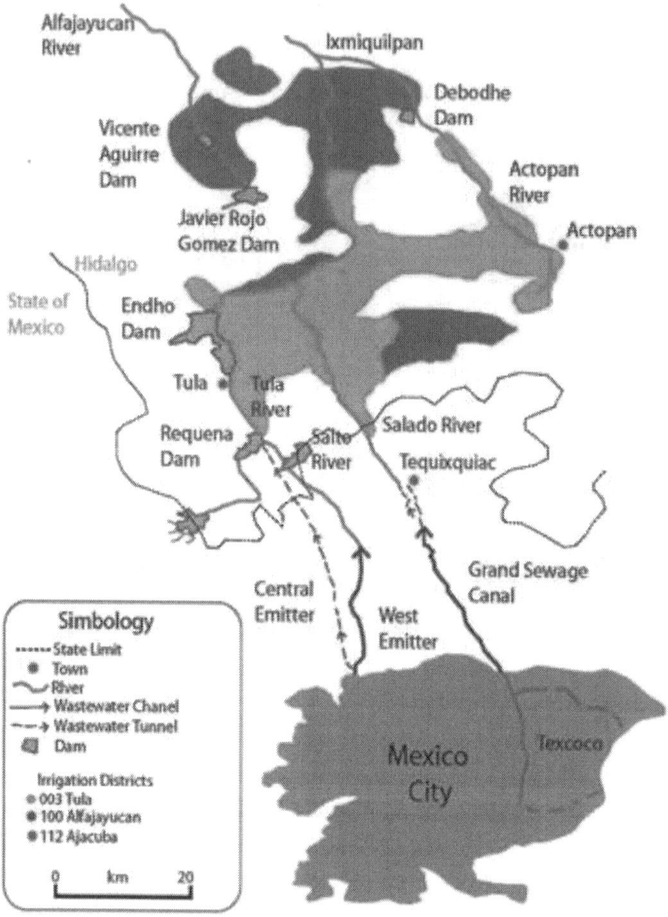

Figure 1. Wastewater transport and use in the Tula Valley.

As wastewater generation gradually increased, the irrigated area grew from 10,000 ha in 1920, to 80,888 ha today (Figure 2). This means that about 1,350 Mm^3/y are transported a distance of 98 km to the Tula Valley. Once in the Valley, wastewater is distributed through a complex hydraulic system that includes six storage dams (combined capacity: 347 Mm^3), 323 km of main distribution canals, 264 km of lateral distribution canals, and 101 km of agricultural drains. Year-round availability of wastewater, as well as its organic matter and nutrient content which act as fertilizers, allowed the development of three important Irrigation Districts (ID 003, 100, and 112) that today cover more than 85 thousand hectares [15].

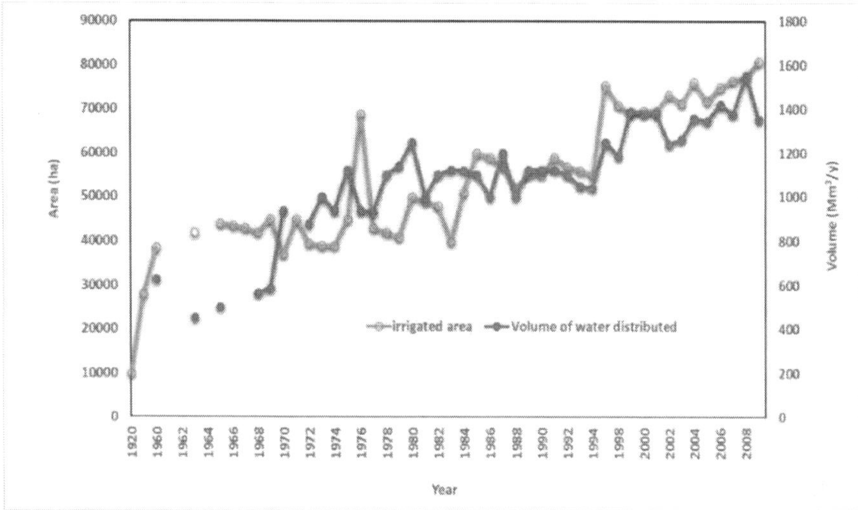

Figure 2. Growth of area irrigated and volume of water distributed since 1920 (adapted from [16], with data from [15]).

Water Reuse in the Tula Valley

The gradual growth of the productive area in the Tula Valley allowed the differentiation of regions with various irrigation ages (Figure 3). Irrigation was first performed in the south-central area of the Valley [14]. Subsequently, when the City expanded and had to dispose of a larger amount of combined wastewater, the irrigation area grew towards the north and the east following the Tula River. Subsequent expansions of the Irrigation Districts carried wastewater to the east and southeast (ID 100), where reuse had started by 1970 [16].

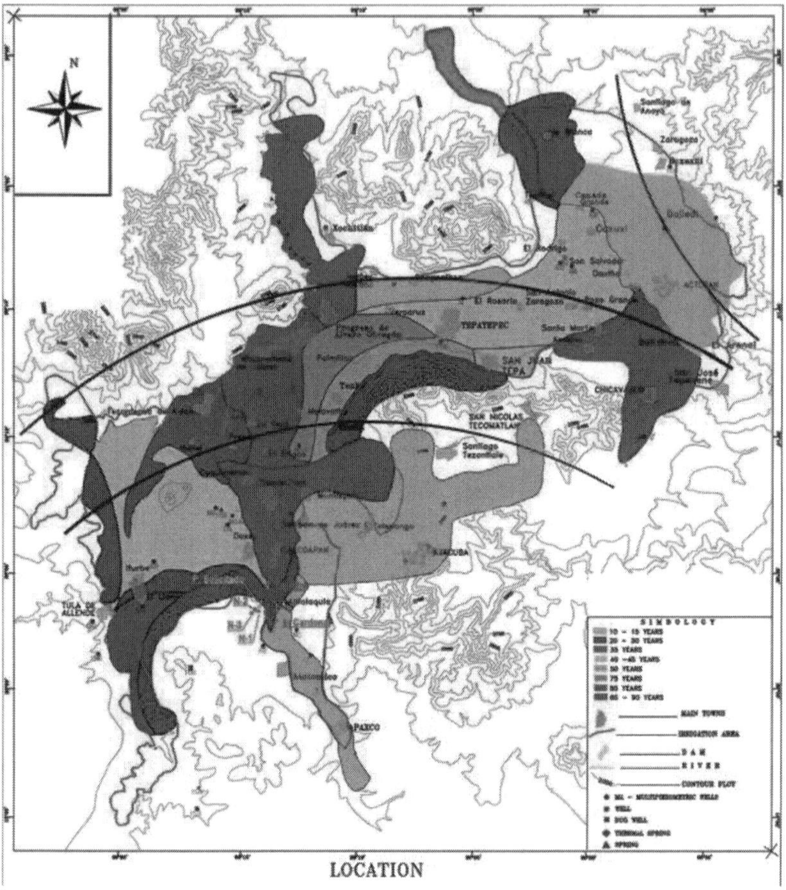

Figure 3. Areas under irrigation for different lengths of time (adapted from [16]).

Recently, wastewater was taken to the far east of the Valley, close to the Ajacuba Municipality (ID 112). As a result, the quality of irrigation water exhibits spatial and temporal variations. For example, in the southern part of the Valley, approximately 10,000 ha are irrigated with raw wastewater, while about 35,000 ha in the central and eastern regions receive diluted wastewater (80% wastewater and 20% river water/precipitation from the Taxhimay and Requena dams) with a different composition.

On the other hand, 25,000 ha located in the far west region of the Valley are irrigated with partially treated wastewater after being stored in the Endho dam which acts as a large settling tank with a hydraulic retention time of up to three months. Finally, the northern areas use well water or return flows and are mainly utilized for growing vegetables and fruits [17].

Water Quality

Microbiological Indicators

The potential risk of bacterial, viral and parasitic diseases that can be transmitted through the human-water-soil-crop-human cycle constitutes the greatest problem associated with the use of wastewater in agriculture, for human consumption or for other uses. Thus the study of their removal is especially relevant in developing countries where they represent the higher risk of disease.

The presence of microorganisms can be a serious cause for concern. For example bacteriophages (viruses) are resistant in the environment, have been shown to have the capacity to penetrate and reach confined aquifers, and have low infective doses [7]. Cysts of *Giardia lamblia* can survive and remain active for months or even years, are resistant to chlorination, have a low infective dose and can have serious consequences for vulnerable individuals [18]. Helminth eggs also survive for long periods of time in the environment, have a low infective dose and their incidence in Mexico is very high [19]. Faecal coliforms are a universally accepted indicator of faecal contamination and have been found to migrate through soil [20].

Wastewater from the MAVM reaches the distribution system at the Tula Valley without any treatment and thus it contains a high concentration of faecal coliforms (between 10^5 to 10^8 colony forming units, CFU/100 mL), *Streptococcus faecalis* (10^7-10^6 CFU/100 mL), *Clostridium perfringens* (10^3 to 10^6CFU/100 mL), somatic bacteriophages (10^2 to 10^6 plaque forming units, PFU/mL), *Giardia* spp. (450 to 10,000 cysts/L), and helminth eggs (1.8 to 23 helminth eggs/L) (Table 1). This quality has shown little temporal variation but the concentrations registered pose a significant health risk in areas under irrigation, according to WHO [4].

To evaluate temporal and spatial variability along the Valley, samples from different regions have been taken for microbial analyses during the wet and dry seasons. Zone 1 (south) uses raw wastewater for irrigation and has shown higher concentrations of biological indicators than Zones 2 or 3 which correspond to areas with some sedimentation and dilution (rainfall and groundwater). This variability was due to the nature of the wastewater but there was a large decrease in the *Giardia* spp. content in all zones during the dry season. Helminth eggs showed little variability within the three zones and there was no obvious difference between the wet and dry seasons. However, both *Giardia* spp. and helminth eggs were relatively high in zone 3 at the farther reaches of the distribution system, which suggests a possible direct contribution from local discharges, since population in this zone is relatively high (Figure 4). The concentrations of microorganisms reported are similar to those previously described for untreated wastewater in Mexico [19] and are similar to untreated wastewater from other countries [21, 22], but much greater than concentrations seen in effluents of treatment plants [19].

Table 1. Microbiological quality of raw wastewater and characteristics of selected indicators (values in parenthesis indicate range of concentration).

Parameter	Concentration	Description
Faecal Coliforms, Log(CFU/100 mL)	6.53 ± 0.58 (5.15-7.84)	Traditional indicator of faecal contamination; may indicate the presence of pathogens; behaves similarly to other bacteria under different environmental conditions.
Streptococcus faecalis, Log(CFU/100 mL)	5.15 ± 1.0 (2.04-6.20)	Intestinal bacteria found in the faeces of all warm-blooded mammals; useful for indicating the quality of water for recreational use and also for reuse.
*Clostridium perfringens*spores, Log(CFU/100 mL)	4.70 ± 0.81 (3.00-5.97)	Anaerobic bacteria that have been recently used as an indicator of faecal contamination; they form a resistant spore commonly found in faeces which is more resistant to disinfection and adverse environmental conditions that many pathogens; the presence of vegetative cells indicates recent contamination while spores imply prior contamination.
Somatic bacteriophages, Log (PFU/mL)	3.63 ± 0.77 (2.41-6.41)	Viruses that infect bacteria; used as indicators based on ease of detection by analytical laboratories; the coliphage group has been the best model for enteroviruses, given their similar physical structure and their greater resistance to treatment processes (such as chlorination); it is always present and relatively abundant in wastewater; easily detected over a short time period (24 h).

Giardia spp., Cysts/L	2,231 ± 231 (450-10,000)	Pathogenic protozoa; single celled organisms that develop in two ways: as trophozoites, and as cysts; Infection results from the consumption of a mature cyst that is resistant to the gastric juices; the trophozoites can become cysts again in a process that is apparently aided by unideal luminal conditions, and they are then expelled in the faeces of persons with symptoms of the illness or with no apparent symptoms. Cysts can survive and remain active for weeks, months or even for periods of up to 7 years; capable of forming cysts (resistant structures) under adverse conditions, this organism represents a serious health risk.
Helminth eggs, eggs/L	13.1±6.2 (1.8-23.0)	Group of parasitic and free-living worms of various sizes and shapes; they cause mechanical deterioration, tissue damage, toxic effects, and blood loss; intestinal parasites cause anaemic malnutrition and/or delayed growth; they present high resistance to chemical and physical conditions and thus are considered the most resistant form of parasites; they have the ability to survive long periods of time in biosolids and soil (up to 6 years after their initial application).

Although it was expected that concentrations of microorganisms would be lower in the wet season due to rainfall dilution, this happens intermittently as the wet season is characterized by irregular intense storms. An additional issue is that the wastewater is mixed with stored rainwater and wastewater depending on the amount of water required by farmers and thus its quality exhibits some variability.

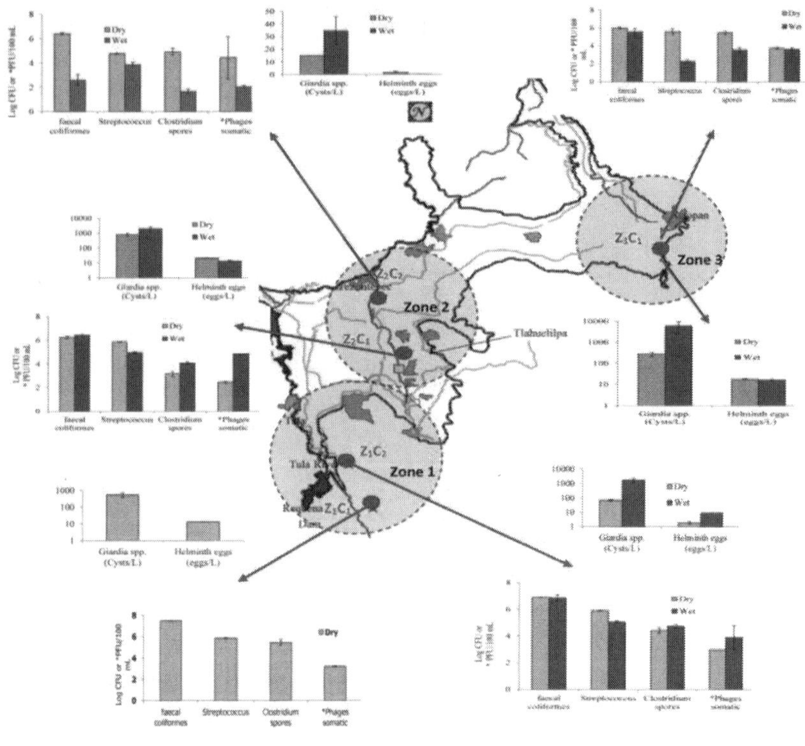

Figure 4. Concentration of selected microorganisms in different zones of the Tula Valley (Adapted from [10] and [23])

Emerging Pollutants

Emerging pollutants are defined as those unregulated pollutants that may be controlled by future regulations depending on their potential effects on health and ecosystems [24]. Advances in analytical techniques for their detection and an increased understanding of their effects on public health and the environment, has increased the need to establish discharge limits worldwide. During the last two decades, the study of emerging pollutants (such as pharmaceuticals, personal care products, and endocrine disruptors included as additives of gasoline, plastics, or detergents) in the environment has gained scientific attention. Many of these studies have been performed in aquatic ecosystems in the United States of America and Europe [24]. However, for most of the emerging pollutants there are still insufficient data to indicate their risk and ecotoxicity and thus their effects on human health or aquatic organisms remains unknown. Many of these compounds reach the environment through wastewater and they may be removed in wastewater facilities, however, some of them remain in treated effluents and enter the environment, eventually reaching groundwater [25, 26].

Initial reports on pharmaceuticals in wastewater were published in the United States in the 1970's [27], however, even though their presence has been related to fish toxicity, little attention has been given to them [28, 29]. Considering that developed countries have detected emerging pollutants in water and wastewater, even after treatment, their presence in wastewater from the Metropolitan Area of the Valley of Mexico was clearly expected. Furthermore, the fact that many prescription drugs in developed countries are considered over-the-counter medications in Mexico, there is a higher probability of finding their active ingredients in wastewater.

On the other hand, agricultural irrigation favors the transport of pollutants, such as pesticides, from soil to groundwater. Several factors influence their incidence in groundwater, including soil permeability, the depth of the unsaturated zone, geological composition, as well as the solubility, partition ($K_{o/w}$), and dissociation (pKa) values of the pollutant. These conditions may promote their adsorption to soil particles or their leaching to groundwater. It has been reported that wastewater irrigation, treated or untreated, increases the concentration of pesticides in groundwater [30].

As mentioned previously, the presence of emerging pollutants was expected in wastewater from the MAVM, and thus at the Tula Valley, and was confirmed by different studies. Some of them [31] analyzed wastewater samples and found a variety of compounds with the exception of clofibric acid. Concentrations found were similar to those reported for wastewater samples taken at treatment plants (Table 2), but higher than those reported for their effluents [32, 33]. The concentrations of naproxen, a widely use pharmaceutical in Mexico, as well as salicylic acid were higher than those reported in other countries, while ibuprofen, diclofenac, are variable in comparison. Other compounds like 4-nonylphenol and diethylhexyl phthalate were the predominant endocrine disruptors in wastewater. Levels of Triclosan, bisphenol-A, butylbencylphtalate, estrone, and 17β-estradiol were similar to those reported elsewhere [33].

It was observed that dilution occurred during the rainy season when reported concentrations of endocrine disruptors were about half of those during the dry season. It should be noted that wastewater at the Tula Valley includes precipitation from the MAVM. Nonetheless, acid pharmaceuticals remain similar in both seasons which may suggest another phenomenon.

Siemens *et al.* [39] concluded that six acid and five basic pharmaceuticals measured in raw wastewater, irrigation water, and spring water, were reduced in concentration along the wastewater distribution canals. However, the authors suggest that acid compounds may permeate clay soils and exhibit poor removal compared to basic or neutral chemicals. Nevertheless, none of the reported compounds were found in spring water.

Table 2. Concentration of pharmaceutical compounds in wastewater.

Compound	Germany[a,c] ng/L	Spain[d] ng/L	Finland[e] ng/L	Brazil[b,c] ng/L	Mexico[f] ng/L
Ibuprofen	3,400	–	13,100	3,300	2,500
Diclofenac	2,000	2,600–5,700	–	800	1,607
Ketoprofen	300	–	2,000	500	447
Salicylic acid	340-5,400	–	–	–	29,867
Naproxen	440	1,800–4,600	4,900	600	13,620

[i] - a[34, 35] b [36]; c[25]; d[37]; e[38]; f[31].

Natural Attenuation of Basic Parameters, Organic Pollutants and Pathogens in Soil

In the Tula Valley, known as the largest case of indirect wastewater reuse for human consumption in the world, natural soil aquifer treatment (SAT) has been taken place for more than 100 years, recharging the aquifer and acting as a barrier to prevent contaminants from entering it. The local aquifer is being recharged at a rate of 25 m^3/s due to the infiltration of untreated wastewater from unlined irrigation channels, storage dams, and excess water used for irrigation (flood irrigation practice; [12]). During percolation, natural soil infiltration occurs through unsaturated soil; as the effluent moves through the soil and the aquifer, it can undergo significant quality improvements through physical, chemical and biological processes. Non-intentional natural SAT has treatment benefits in the unsaturated zone in the Tula Valley. It acts as a natural filter, and produces groundwater of acceptable quality due to the characteristics of the soil-aquifer structure, the residence time, and the history of the complex geohydrological system. Table 3showed the historical data of wastewater and aquifer water quality variability obtained from studies performed at the site. It indicates that parameters related with salinity behavior are present in groundwater, in addition to parameters such as nitrate, and solids in dissolved phase. In contrast, the percentage removal of pathogenic organisms and emerging pollutants through unsaturated soil is greater than 50%.

There is evidence that during infiltration the vast majority of microorganisms are retained in the first few centimeters of soil [40], however their potential fate is influenced by size ranges of microorganisms (20-80 μm for helminth eggs, 1 μm for bacteria, 4-12 μm for protozoa), type of soil, and even soil organic matter content for bacteriophages (20 to 200 nm), and therefore for viruses; thus the migration of microorganisms through soil does not always allow complete attenuation.

Table 3. Water quality variability for wastewater and the Tula Valley aquifer and natural removal percentages (SAT)

Parameters	Units	Wastewater (Min-Max)	Aquifer (Min-Max)	% removal (mean)
Basic analysis				
Total Dissolved Solids, TDS	mg/L	409-1123	546 -1586	-55.5
Alkalinity	mg/L	324-600	256-748	-7.8
Nitrate (as N)	mg/L	0.02-0.47	0.0-32	-8014
Salinity related analysis				
Electrical conductivity	μS/cm	1734-3000	1755 -4187	-18
Redox potential	mV	-14	-68.1	-38.7
Sodium	mg/L	56-215	103 -361	-38.3
Potassium	mg/L	19-13190	10.5 -107	98
Hardness Ca	mgCaCO$_3$/L	90-131	70-787	-218.4
Hardness Mg	mgCaCO$_3$/L	Dec-51	17 -301	-65.7
Total hardness		122-172	114-1006	-181.8

Bicarbonate	mg/L	270-504	212-652	-12.8
Sulfate	mg/L	53-2492	190 -3025	-15
Microbiological analysis				
Total Bacteria count (37 °C)	Log CFU/100 mL	6.4-7.7	1.3-4.2	57.6
Faecal coliforms	Log CFU/100 mL	5.6-7.5	0.0-2.9	70.8
Enterococci	Log CFU/100 mL	2.3-5.9	0.0-1.9	79.3
Clostridium spores	Log CFU/100 mL	3.2-5.6	0.0-2.6	61.1
Somatic Bacteriophages	Log PFU/100 mL	2.5-4.9	0.0-2.8	52.6
Giardia spp.	Cysts/L	70.3-3233	0.0-600	89.4
Helminth egg	Ova/L	1.9-21.9	0.0-2.9	96
Emerging pollutants				
Clofibric acid	ng/L	< LOD	< LOD-0.39	–
Ibuprofen	ng/L	1325-4700	0.05-1.46	99.99
Salicylic acid	ng/L	13580-72979	0.02-27.7	99.98
2,4-D	ng/L	295-2641	< LOD-0.48	99.99
Gemfibrozil	ng/L	640-750	< LOD-0.1	99.99
Naproxen	ng/L	5861-16336	0.04-3.266	100
Ketoprofen	ng/L	82-500	0.02-0.83	99.89
Diclofenac	ng/L	1240-3424	< LOD-3.75	99.98
4-nonylphenol	ng/L	6970-38130	0.81-67.6	99.93
Pentachlorophenol	ng/L	40-110	< LOD-0.33	99.81
Triclosan	ng/L	360-2880	< LOD-22.6	99.83
Bisphenol-A	ng/L	700-6230	0.02-153	99.54
Butil-bencilphthalate	ng/L	125-2959	0.03-308	97.78
Bis-2 ethyl(hexyl)phthalate	ng/L	4664-70200	3.07-933	99.97
Estrone	ng/L	14-100	< LOD-0.24	99.92
17β-estradiol	ng/L	6.8-22	< LOD-0.06	99.9
17α-etinilestradiol	ng/L	<LOD	< LOD-0.05	–
Carbamazepine	ng/L	200-275	0.14-193	85.89

[i] - Adapted from [23, 31, and 39] (ND: Not detected LOD: Lower limit of detection).

Data recorded over many years regarding the attenuation of emerging pollutants through soil show a mean reduction of 86 percent of the chemical load in the wastewater used for irrigation (Table 3). The main attenuation processes such as adsorption and biodegradation have been confirmed in experimental studies with local soil samples, and enriched samples [23]. The most important result of this research is the observation that large amounts of suspended and dissolved organic matter in soils [41], and even in raw wastewater, improve the adsorption of emerging pollutants in soils, and therefore their final low content in groundwater; this verifies previously published results with regard to the role of natural SAT in removing pollutants [26, 42]. In fact, the sorption behavior of three pharmaceuticals (naproxen, carbamazepine, and triclosan), a plasticizer Bis-2 ethyl(hexyl)phthalate, and the surfactant metabolite 4-nonylphenol in wastewater irrigated soil, was analyzed for different soil depths sampled from the Tula Valley, showing that the potential migration of these compounds to the aquifer depends on the physical and chemical characteristics of soil, such as organic matter content, and clay/sand percentages [43, 44].

Other studies have shown that heavy metals have accumulated in soil in the Tula Valley (for example cadmium, nickel, and lead) and their retention is associated with the length of time that wastewater has been used for irrigation [45].

It should be mentioned that a large wastewater treatment plant is currently under construction to treat the MAVM's wastewater and reduce pollutant load to the Tula Valley. This plant, Atotonilco, is designed to treat 35 m^3/s with a combined process (23 m^3/s with a biological process and 12 m^3/s with a physicochemical process) that may partially remove organic matter and the bulk of pathogens. However, the effect on soil attenuation must be evaluated as Gibson *et al.* [41] suggest that a reduction in the amount of organic matter that gets into irrigated soils may affect sorption processes, which could have an impact on the removal of emerging pollutants, pathogens, and heavy metals.

Impacts of Wastewater Irrigation in the Mezquital Valley

Reuse in the Tula Valley has functioned as a discharge route for the closed basin of the MAVM at a relatively low cost. The characteristics of the water, in terms of organic matter and nutrients have significantly increased crop productivity in the Valley and have allowed farmers to harvest up to five crops per year of alfalfa, fodder oats, tomato, barley, and maize. Resulting yields in the Tula Valley are 71 to 150% higher than those obtained with rain-fed agriculture [19]. Thus, the contribution of nutrients represents estimated annual savings of $180 to $200 million USD. To exemplify this, agricultural production for 2011-2012 in the Irrigation Districts of the Tula Valley reached $418 million USD [46].

Since 81% of the main distribution canals as well as 52% of lateral distribution canals are unlined, 80.2 Mm^3/y of conveyed wastewater infiltrates to the aquifer [47]. It is with this infiltration that wastewater quality is improved before reaching the groundwater through processes like adsorption or biodegradation, which depend on contact time and filtering distance. As an example, travel times from the irrigated fields to the springs are estimated to be 3 to 5 days with groundwater velocities of 0.02 to 6.0 m/d [16].

This infiltration has also incidentally recharged the Tula Valley aquifer for more than 100 years, at a recharge rate estimated at 25 m^3/s and equivalent to 13 times the natural recharge [16]. As a consequence, the water table has risen from a depth of 60 meters in the 1950's to 4 meters in the southern part of the Valley and several springs with flows between 40 and 600 L/s have appeared [16]. The amount of available water with a relatively good quality allows some areas to be used to grow vegetables like tomato, lettuce, cabbage, beetroot, cilantro, radish, carrot, spinach, and parsley [16]. At the same time, surface and groundwater produced by wastewater infiltration provide drinking water to approximately 500,000 people after only treatment with chlorine for disinfection [12].

In contrast, wastewater reuse in the Tula Valley has some negative impacts on the local population and the environment. Due to the pathogens and parasites contained in wastewater, the incidence of gastrointestinal diseases has increased by more than 16 times in children living in the irrigation area, compared to children unexposed to wastewater [48]. In addition, microbial and organic pollutants have been detected in the soil matrix of the irrigation areas at various depths [41, 44]. Heavy metals have accumulated in the upper soil layer [49], and Siebe [45] suggests that eventually the retention capacity of the soil might be exceeded with the risk of groundwater pollution. Moreover, the gradual salinization of local soils has caused the loss of more than 2,000 ha of cropland [17].

Conventional Wastewater Treatment Processes, Control of Micropollutants, and Hazardous Substances in Water

Even though the soil filters a large amount of pollutants from wastewater with removals above 90% [12], some studies have demonstrated that in some areas the filtered water still contains bacteria (coliforms and enterococci) and viruses [40, 50]. Filtered water presents a high concentration of salts (indicated by sodium, sulfates, carbonates, bicarbonates, chloride, calcium hardness, electrical conductivity, and potassium), and furthermore, total dissolved solids and nitrates exceed the limits set by the Mexican regulatory authorities (1,000 and 10 mg/L, respectively).

At the same time, several studies [23, 31, 40, 41, 43, 44] report the presence of emerging pollutants in springs and wells, which poses a health risk to local population (approximately 500,000 inhabitants) that use the water for human consumption (Figure 5). These pollutants exhibit low removal by conventional treatment processes and thus alternative technologies, such as membrane filtration (nanofiltration or reverse osmosis), have been proposed [51]. Different studies have demonstrated that nanofiltration removes emerging pollutants, including those with low molecular weight, in particular pesticides [52], pharmaceuticals [51], and endocrine disruptors [52].

As a result, to improve the quality of spring water in the Tula Valley, a pilot plant with a nanofiltration membrane was installed in the Cerro Colorado Spring (Figure 5) [53]; The pilot plant (11.4 m^3/d) operated for 800 hours and included a pumping system, a prefilter, a softening unit for reducing potential scaling, and a nanofiltration module (Figure 6). The nanofiltration (NF) membrane was selected after laboratory trials on a pilot cell. The pilot plant had the instrumentation required to measure flow, pH, TDS, as well as pressure along the system. Recovery (permeate) was maintained at more than 66 % of the influent.

Figure 5. Cerro Colorado Spring at the Tula Valley.

Nanofiltration demonstrated its effectiveness for removing organic matter (92% of total organic carbon), salinity (60% of total dissolved solids and 75% of electrical conductivity) and 100% for selected microorganisms (faecal coliforms, faecal streptococci, *Salmonella* spp., *Clostridium perfringens*, *Giardia* spp., and bacteriophages). With respect to emerging pollutants, membrane selectivity varied from 5 to 6% for salicylic acid and nonylphenol up to more than 75% for gemfibrozil, butylbenzyl phthalate, carbamazepine, and diclofenac. Treated water (permeate) met the limits for drinking water and thus it could be considered suitable for human consumption.

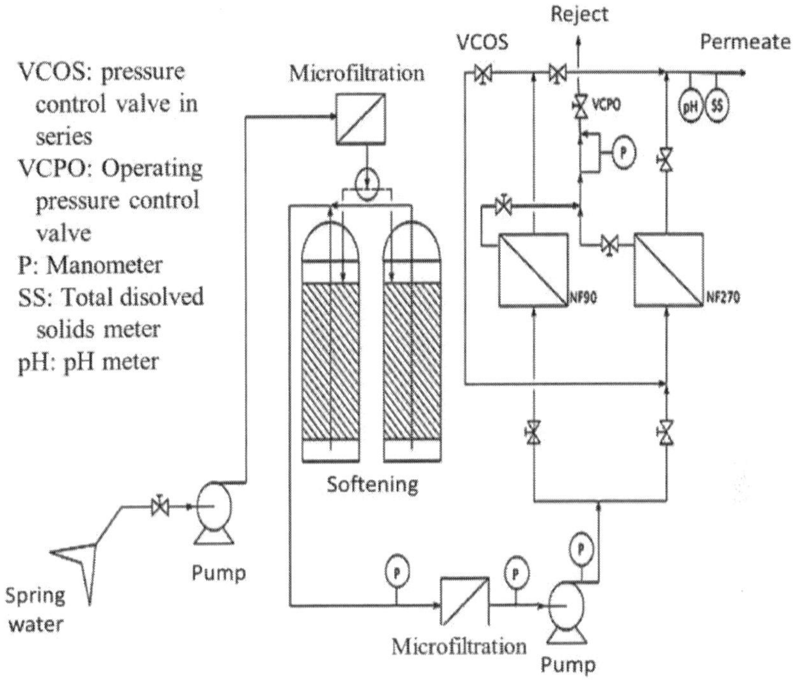

Figure 6. Flow diagram of the nanofiltration pilot plant at the Cerro Colorado Spring, Tula Valley.

Based on such studies, filtered water from the Tula Valley is being considered as a potential source for supplying drinking water for Mexico City and studies have determined that 6.5 m^3/s may be sourced from local groundwater [54].

FORTHCOMING EXPECTATIONS AND RECOMMENDATIONS

Findings confirmed by 15 years of research into the complex Tula Valley hydrological and hydraulic system, are:

- Wastewater reuse for irrigation has proven to be an alternative in the water-scarce region of Tula Valley.

- Agricultural irrigation in the Tula Valley using raw wastewater from the Metropolitan Area of the Valley of Mexico has been performed for more than 100 years and several studies have demonstrated that complex phenomena occur.
- One of these is the natural SAT in the saturated soil zone, which even occurs today after a century of irrigation, mainly in terms of the retention of typical pathogenic organisms and emerging contaminants in wastewater of urban origin.
- The observed level of natural removal through filtration by soil exceeds the expected removal levels from conventional wastewater treatment, not only for emerging contaminants, but also for microorganisms that are difficult to remove by chlorine disinfection.
- Non-intentional aquifer recharge with water purification by natural infiltration, achieved through agricultural reuse of untreated wastewater, has provided new water sources for the local population with the potential to supply water for human consumption to the large city from which comes the reused wastewater originates.
- Advanced treatment studies conducted by our research group have shown that it is possible to treat water from aquifer for safe water supply.

There is no doubt about the benefits of wastewater reuse for irrigation, but some potential impacts should be considered:

- Even though the soil matrix retains pollutants, the limit of the soil retention capacity remains uncertain, largely for metals and emerging compounds.
- The content of organic matter in soil has been a key factor in the natural attenuation of water pollution, however, the impact of future changes in the quality of water for irrigation purposes in Tula Valley, may affect removal capacity.
- Soon, the Atotonilco WWTP will supply wastewater to the irrigation canals with a lower content of organic matter and nutrients. These are factors which have previously favored the natural process of the SAT and the development of agricultural activity in the area. These reductions may affect complex physicochemical processes in the future, and may have impacts on soil and groundwater quality that cannot be predicted.
- The use of treated wastewater for irrigation, will allow the growing of crops normally eaten raw, which up till now has been forbidden by

national legislation. Other benefits will be the reduction of direct exposure of farmers to pollutants in untreated water.

- Finally, with regard to national and international legislation for irrigation, it is evident that local studies are required for establishing permissible limits. This is because existing guidelines require wastewater treatment levels which are economically onerous, mainly for developing countries, and moreover, do not include limits for other pollutants present in urban wastewater (pharmaceuticals and personal care use products). Wastewater will become more and more attractive for agricultural reuse, given the current and future problems of water scarcity.

REFERENCES

1. FAO. Monitoring agricultural water use at country level. Experiences of a pilot project in Benin and Ethiopia. Land and Water Discussion Paper 9. ISSN 1729-0554. 2011.
2. Hamilton A.J., Mebalds M., Aldaoud R., Heath, M. A survey of physical, agrochemical and microbial characteristics of waste-water from the carrot washing process: implications for re-use and environmental discharge. Journal of Vegetable Science. 2005. 11, 57–72.
3. Peasey A., Blumenthal U., Mara D., Ruiz-Palacios G.A. Review of Policy and Standards for Wastewater Reuse in Agriculture: A Latin American Perspective. London School of Hygiene & Tropical Medicine. 2000.
4. World Health Organization. Guidelines for the Safe Use of Wastewater, Excreta, and Greywater. Vol. 2, Wastewater Use in agriculture. Geneva, Switzerland. 2006.
5. Blumenthal, U., Peasey, A. Critical Review of Epidemiological Evidence of the Health Effects of Wastewater and Excreta Use in Agriculture. London School of Hygiene and Tropical Medicine. 2002.
6. Page, D., Dillon, P., Toze, S., Bixio, D., Genthe, B., Jiménez Cisneros, B.E., Wintgens, T. Valuing the subsurface pathogen treatment barrier in water recycling via aquifers for drinking supplies. Water Research. 2010;44(6) 1841-1852.
7. Asano, T., editor. Wastewater Reclamation and Reuse. Water Quality Management Library Vol. 10. Technomic Publishing Inc., Lancaster, Pennsylvania, USA. 1998.
8. FAO (1992) Wastewater treatment and use in agriculture-FAO irrigation and drainage paper 47 http://www.fao.org/docrep/T0551E/t0551e00.htm

9. Medaware. Development of Tools and Guidelines for the Promotion of the Sustainable Urban Wastewater Treatment and Reuse in the Agricultural Production in the Mediterranean Countries. European Commission Euro-Mediterranean-Partnership. Technical Guidelines on Wastewater Utilization. ME8/AIDCO/2001/0151/59341-P033. 2005.
10. Jimenez, B. and Asano, T. Water reuse: an international survey of current practices, issues and needs. IWA Publishing. ISBN: 1843390892, 650 pp. 2008.
11. Jiménez, B. Treatment technology and standards for agricultural wastewater reuse: a Case Study in Mexico. Irrigation and Drainage. 2005;54(1) 23-25.
12. Jiménez, B., Chávez, A. Quality assessment of an aquifer recharged with wastewater for its potential use as drinking source: "El Mezquital Valley" case. Water Science and Technology. 2004;50(2) 269-276.
13. Domínguez, R. El sistema principal de drenaje del Área Metropolitana del Valle de México. Series Instituto de Ingeniería. CI-22, 34 p. In Spanish. 2001.
14. Cruz Campa, S. Rehabilitación Integral del Distrito de Riego 03, Tula, Hidalgo. Thesis, Escuela Nacional de Agricultura, Chapingo, México. In Spanish. 1965.
15. CONAGUA. Comisión Nacional del Agua. Estadísticas del agua en México, 2010. Secretaría del Medio Ambiente y Recursos Naturales. México D.F. In Spanish. 2010.
16. British Geological Survey-Comisión Nacional del Agua (1998). London School of hygiene and tropical Medicine and University of Birmingham. Impact of wastewater reuse on groundwater in the Mezquital Valley, Hidalgo State, Mexico. Department for International Development, DFID Final Report. Technical Report WC/98/42. 1998.
17. Siebe C. and Cifuentes E. Environmental impact of wastewater irrigation in central Mexico: An overview. International Journal of Environmental Health Research. 1995; 5: 161-173.
18. Brusseau, M., Oleen, J., Santamaria, J., Cheng, L., Orosz-Coghlan, P., Chetochine, A., Blandford, W., Rykwalder, P., Gerba, C. Transport of*Microsporidium Encephalitozoon* Intestinales Spores in Sand Porous Media. Water Research. 2005; 39: 3636-3642.
19. Jiménez, B., Chávez, A., Hernández, C. Alternative Wastewater Treatment Intended for Agricultural Use. Water Science and Technology. 1999; 40(4-5) 355-362.
20. Dunsmore, B., Bass, C., Lappin-Scott, H. A Novel Approach to Investigate Biofilm Accumulation and Bacterial Transport in Porous Matrices. Environmental Microbiology. 2004; 6(2) 183-187.
21. Blumenthal, U., Mara, D., Ayres, R., Cifuentes, E., Peasey, A., Scott, R., Lee, D., Ruiz-Palacios, G. Evaluation of the WHO nematode egg guidelines for

restricted and unrestricted irrigation. Water Science and Technology. 1996; 33:277-283.

22. Modella, P., Blanco, I., Ansola, G., de Luis, E. Removal of Wastewater Pathogen Indicators in a Constructed Wetland in Leon, Spain. Ecological Engineering. 2008; 33(3-4) 252-257.
23. Jiménez, B., Chavez, A., Gibson, R., Maya, C. Unplanned aquifer recharge in El Mezquital/Tula Valley, México. In: Kazner, C., Wintgens, T., Dillon, P. (eds.) Water reclamation technologies for safe managed aquifer recharge. IWA Publishing. London, 2012. 141-143.
24. Barceló, D. Emerging pollutants in water analysis. Trends in Analytical Chemistry. 2003; 22(10): 685-696.
25. Daughton, C.G. and Ternes, T.A. Pharmaceuticals and Personal Care Products in the Environment: Agents of Subtle Change? Health Perspective. 1999; 107: 907-928.
26. Kümmerer, K. (ed). Pharmaceuticals in the Environment. Source, Fate, Effects and Risks. 2nd edition. Springer. 2004.
27. Tabak, H.H. and Bunch, R.L. Steroid hormones as water pollutants. I. Metabolism of natural and synthetic ovulation inhibiting hormones by microorganisms of activated sludge and primary settled sewage. Developments in Industrial Microbiology. (1970). 11, 367–376
28. Kramer, V.J., Miles-Richardson, S., Pierens, S.L., Giesy, J.P. Reproductive impairment and induction of alkaline-labile phosphate, a biomarker of estrogen exposure, in fathead minnows (*Pimephales promelas*) exposed to waterborne 17β-estradiol. Aquatic Toxicology. 1998; 40(4) 335–360.
29. Snyder, S.A., Villeneuve, D.L., Snyder, E.M., Giesy, J.P. Identification and quantification of estrogen receptor agonists in wastewater effluents. Environmental Science and Technology. 2001; 35(18) 3620–3625.
30. Bixio, D., Thoeye, C., De Koning, J., Joksimovic, D., Savic, D., Wintgens, T., Melin, T. Wastewater reuse in Europe. Desalination. 2006; 187: 89-101.
31. Gibson, R., Becerril-Bravo, E., Silva-Castro, V., Jiménez-Cisneros, B. Determination of acidic pharmaceuticals and potential endocrine disrupting compounds in wastewaters and spring waters by selective elution and analysis by gas chromatography–mass spectrometry. Journal of Chromatography A. 2007; 1169: 31–39.
32. Verenitch, S.S., Lowe, C.J., Mazumder, A. Determination of acidic drugs and caffeine in municipal wastewaters and receiving waters by gas chromatography–ion trap tandem mass spectrometry. Journal of Chromatography A. 2006; 1116: 193-203.
33. Lee, H.B., Peart, T.E., Svoboda, M.L. Determination of endocrine-disrupting phenols, acidic pharmaceuticals, and personal-care products in sewage by

solid-phase extraction and gas chromatography-mass spectrometry. Journal of Chromatography A. 2005; 1094: 122-129.

34. Heberer, T., Fuhrmann, B., Schmidt-Baumler, K., Tsipi, D., Koutsouba, V., and Hiskia, A. Occurrence of pharmaceutical residues in sewage, river, ground and drinking water in Greece and Berlin (Germany). In: Daughton, CGTJ-L, (ed.) Pharmaceuticals and Personal Care Products in the Environment: Scientific and Regulatory Issues. American Chemical Society. Washington, DC. 70-83. 2001.
35. Farré, M., Ferrer, E., Ginebreda, A., Figueras, M., Olivella, L., Tirapu, L., Villanova, M. and Barcelo, D. Determination of drugs in surface water and wastewater samples by liquid chromatogrphy-mass spectrometry: methods and preliminary results including toxicity studies with *Vibrio fishery*. Journal of Chromatography A. 2001; 938: 187-197.
36. Stumpf, M., Ternes, T.A., Wilken, R.D., Rodrigues, S.V., Baumann, W. Polar drugs residues in sewage and natural waters in the state of Río de Janeiro, Brazil. Science of the Total Environment. 1999; 225(1-2) 135-141.
37. Carballa, M., Omil, F., Lema, J.M., Llompart, M., García-Jares, C., Rodríguez, I., Gómez, M., Ternes, T. Behavior of pharmaceuticals, cosmetics and hormones in a sewage treatment plant. Water Research. 2004; 38(12) 2918-2926.
38. Linqvist, N., Tuhkanen, T., Kronberg, L. Occurrence of acidic pharmaceuticals in raw and treated sewage and in receiving waters. Water Research. 2005; 39(11) 2219-2228.
39. Siemens, J., Huschek, G., Siebe, C., Kaupenjohann, M. Concentrations and mobility of human pharmaceuticals in the world's largest wastewater irrigation system, Mexico City–Mezquital Valley. Water Research. 2008; 42(8-9) 2124 –2134.
40. Chávez, A., Maya, C., Gibson, R., Jiménez, B. The removal of microorganisms and organic micropollutants from wastewater during infiltration to aquifers after irrigation of farmland in the Tula Valley, Mexico. Environmental Pollution. 2011; 159(5) 1354-1362.
41. Gibson, R., Durán-Álvarez, J.C., León, K., Chávez, A., Jiménez, B. Accumulation and leaching potential of some pharmaceuticals and potential endocrine disruptors in soils irrigated with wastewater in the Tula Valley, Mexico. Chemosphere. 2010; 81(1) 1437–1445.
42. Birkett, J.W.; Llester, J.N. Endocrine disrupters in wastewater and sludge treatment processes. CRC Press. 2002.
43. Murillo, R., Durán-Álvarez, J.C., Prado-Pano, B., Jiménez-Cisneros, B. Mobility of 4-nonylphenol and di(2-ethylhexyl) phthalate in three agricultural soils irrigated with untreated wastewater. Water Science and Technology. 2012; 66(2) 292-298.

44. Durán-Álvarez, J.C., Prado-Pano, B., Jiménez-Cisneros, B. Sorption and desorption of carbamazepine, naproxen and triclosan in a soil irrigated with raw wastewater: Estimation of the sorption parameters by considering the initial mass of the compounds in the soil. Chemosphere. 2012; 88(1) 84-90.
45. Siebe, C., Fischer, W.R. Effect of long-term irrigation with untreated sewage effluents on soil properties and heavy metal adsorption of Leptosols and Vertisols in Central Mexico. Z. Pflanzenernaehr. Bodenk. 1996; 159(4) 357-364.
46. CONAGUA. Estadísticas agrícolas de los Distritos de Riego. Año agrícola 2011-2012. 2013. http://www.conagua.gob.mx/CONAGUA07/Publicaciones/Publicaciones/SGIH-4-13.pdf (accessed 26 September 2014).
47. Lesser, L., Lesser, J.M, Arellano-Islas, S., González-Posadas, D. Balance hídrico y calidad del agua subterránea en el acuífero del Valle del Mezquital, México central. Revista Mexicana de Ciencias Geológicas. 2011; 28(3) 323-336. In Spanish.
48. Cifuentes, E. The epidemiology of enteric infections in agricultural communities exposed to wastewater irrigation: Perspectives for risk control. International Journal of Environment and Health. 1998; 8(3) 203-213.
49. Herre, A.; Siebe, Ch. Effects of irrigation water quality on organic matter, Cd, and Cu mobility in soils of central Mexico. Water Science and Technology. 2004; 50(2) 277-284.
50. Gallegos, E., Warren, A., Robles, E., Campoy, E., Calderon, A., Sainz, M. G., Bonilla, P., Escolero, O. The Effects of Wastewater Irrigation on Groundwater Quality in Mexico. Water Science and Technology. 1999; 40(2) 45–52.
51. Kimura, K., Amy, G., Drewes, J., Watanabe, Y. Adsorption of hydrophobic compounds onto NF/RO membranes: an artifact leading to overestimation of rejection. Journal of Membrane Science. 2003; 221(1-2) 89–101.
52. Zhang, Y., Causserand, C., Aimar, P., Cravedi, J.P. Removal of bisphenol A by a nanofiltration membrane in view of drinking water production. Water Research. 2006; 40(20) 3793-3799.
53. Chávez, A., Torner, F., Barrios, J.A., Maya, C. Lucario, E., Jiménez, B. Membrane process for spring water treatment in the Tula Valley. Comparative microbiology assessment. Water Science and Technology: Water Supply. In press.
54. CONAGUA (2014) Strategic projects: water supply, sewerage and sanitation. http://www.conagua.gob.mx/english07/publications/StrategicProjects.pdf (accessed 22 September 2014)

CITATION

I. Navarro, A. Chavez, J.A. Barrios, C. Maya, E. Becerril, S. Lucario and B. Jimenez (2015). Wastewater Reuse for Irrigation — Practices, Safe Reuse and Perspectives, Irrigation and Drainage - Sustainable Strategies and Systems, Dr. Muhammad Salik Javaid (Ed.), ISBN: 978-953-51-2123-7, InTech, DOI: 10.5772/59361.

CHAPTER 5

Measurement and Assessment of Water Resources Carrying Capacity in Henan Province, China

Ming Dou, Jun-xia Ma, Gui-qiu Li, Qi-ting Zuo

College of Water Conservancy and Environment, Zhengzhou University, Zhengzhou 450001, PR China

ABSTRACT

As demands on limited water resources intensify, concerns are being raised about water resources carrying capacity (WRCC), which is defined as the maximum sustainable socioeconomic scale that can be supported by available water resources and while maintaining defined environmental conditions. This paper proposes a distributed quantitative model for WRCC, based on the principles of optimization, and considering hydro-economic interaction, water supply, water quality, and socioeconomic development constraints. With the model, the WRCCs of 60 subregions in Henan Province were determined for different development periods. The results showed that the water resources carrying level of Henan Province was suitably loaded in 2010, but that the province would be mildly overloaded in 2030 with respect to the socioeconomic development planning goals. The restricting factors for WRCC included the available water resources, the increasing rate of GDP, the urbanization ratio, the irrigation water utilization coefficient, the industrial water recycling rate, and the wastewater reuse rate, of which the available water resources was the most crucial factor. Because these factors varied temporally and spatially, the trends in predicted WRCC were inconsistent across different subregions and periods.

INTRODUCTION

The concept of carrying capacity is rooted in demography, biology, and applied ecology (Clarke, 2002). In ecology, carrying capacity is defined as the maximum population of a species that a habitat can support without permanent impairment of the habitat's productivity (Rees, 1997). Water resources carrying capacity (WRCC) is a new concept that has not yet been clearly defined and described. Some researchers consider WRCC to be the capacity of water resources to sustain a society at a defined good standard of living, while others consider it the threshold level of water resources at which an environment is capable of supporting the activities of human beings (Seidl and Tisdell, 1999 and Li et al., 2000). Internationally, not many breakthroughs have been achieved in the WRCC research; the topic has only been considered briefly in theories of sustainable development (Ofoezie, 2002). Some scholars have used terms such as sustainable water utilization, the ecological limits of water resources, or the natural system limits of water resources to express the meaning of WRCC (Hunter 1998 and Falkenmark and Lundqvist, 1998). Studies focusing exclusively on WRCC have primarily been conducted in China. The concept of WRCC was first applied to the Urumqi River Basin in China in 1989 (Shi and Qu, 1992 and Feng et al., 2006). It has been a topic of significant debate since 2001, and represents a new academic frontier (Long et al., 2004).

One definition of WRCC, and the definition used in this study, is the maximum sustainable socioeconomic scale based on available water resources and maintenance of good, defined environmental conditions (Dou et al., 2010). In this concept, the socioeconomic scale is the overall size of a regional socioeconomic system in a certain period, and can be represented by a series of socioeconomic indices (such as total population, urbanization ratio, industrial structure, and grain yield). Good environmental conditions mean a suitable living environment for human beings and the ecological system, in particular good water quality and a healthy aquatic environment. WRCC is an indicator of regional sustainability, and achieving regional sustainability is important because social institutions and ecological functioning are closely linked at this scale (Graymore et al., 2009). Therefore, research on WRCC should be based on two premises: First, it must be possible to sustain the normal operation of a regional socioeconomic system, and as a result researchers must calculate the quantity of water resources required to sustain these social service functions. Second, it is necessary to evaluate the maximum socioeconomic scale that water resources can sustain after meeting the needs of the ecosystem.

Regional carrying capacity depends on water resources. There have been many theoretical studies of carrying capacity based on regional water resources because this concept is most often considered within a larger theoretical context of sustainable development. In particular, severe water shortage problems have forced the Chinese government to initiate a series of studies to determine the carrying capacity based on regional water resources in arid and semi-arid areas, such as western China and the North China Plain (Xia and Zhu, 2002, Dou et al., 2010 and Zai et al., 2011). In recent years, with increasingly serious water pollution, there have even been some studies conducted in eastern China, where water resources are abundant (Liu and Borthwick, 2011 and Liu, 2012). Furthermore, Falkenmark and Lundqvist (1998) have used estimates of the maximum global use of water resources to study how carrying capacity is determined by regional water resources. The National Research Council (NRC) (2002) studied the Florida Keys Basin's carrying capacity in the United States under different land-use scenarios. Lane et al. (2014) offered a Carrying Capacity Dashboard (QUT, 2012) to highlight one way in which some basic resource-based parameters have been utilized. In practice, carrying capacity is often estimated by comparing stress on the environment (e.g., demand of natural resources) against environmental thresholds (e.g., available natural resources) (Clarke, 2002 and Oh et al., 2005).

On the whole, the current studies on WRCC emphasize harmonization of the demands of socioeconomic development with the supply of water resources. Regional socioeconomic systems and water resources systems are often represented using areas such as river basins, which allow researchers to analyze the systems' structures, functions, and processes and determine the WRCC. In China, the regional socioeconomic scale may be determined by the urban population growth rate and economic development goals. Constraints imposed by the availability of water and other natural resources are rarely considered in planning, which may explain why most Chinese cities are facing severe water shortages and experiencing environmental problems (Zhang et al., 2010). Therefore, it is necessary to develop a suitable methodology to effectively describe hydro-economic interaction in highly populated regions and to choose the best strategies to alleviate the conflict between socioeconomic development and water resources exploitation.

Henan Province, China's most populous province and the province with the fifth highest gross domestic product (GDP), has long suffered from an intense conflict between the limited water resources and the rapid growth of water demand. In this study, we developed a method for calculating

Henan's WRCC based on available water resources and relevant water environment protection goals, and analyzed spatial and temporal variations of the WRCC. First, considering the spatial differences of the economic development level and water resources conditions, a distributed hydro-economic model was developed to describe the interaction between the socioeconomic and water resources systems. Second, a WRCC quantification model was developed to identify the maximum sustainable socioeconomic scale based on the hydro-economic interaction relationship and a series of constraint conditions. Finally, on the basis of the models, Henan Province's WRCC was calculated for different development periods, and the change tendency of the water resources carrying level was analyzed.

FRAMEWORK AND METHODOLOGY

Overview

Research on WRCC involves many disciplines, including hydrology, ecology, environmental sciences, economics, sociology, and management science (Zhang et al., 2010). Many methods can be used, of which the most common are trend analysis (Liu, 2012), the fuzzy comprehensive evaluation method (Prato, 2009), system dynamics (Feng et al., 2008 and Dang and Guo, 2012), multi-objective decision-making and analysis (Xu and Cheng, 2000), the large-scale system theory, the optimization method, and the projection pursuit approach (Zhang and Guo, 2006 and Liu and Borthwick, 2011). Trend analysis is based on empirical analysis of some socioeconomic indices under water resources constraints. The fuzzy comprehensive evaluation method is a common assessment method based on a set of index systems. System dynamics can reflect the interaction and feedback mechanism between human activities and the water resources system. Multi-objective decision-making and analysis can obtain the maximum sustainable socioeconomic scale under a series of water environmental and resources constraints. The large-scale system theory can use the idea of decomposition and coordination to solve a large-scale system problem. The optimization method finds the global optimal solution for complex problems based on given criteria. Finally, the projection pursuit approach is a new statistical method solving multi-dimensional socioeconomic and water resources system problems. From a management perspective, the large-scale system theory and the optimization method are most appropriate because the former shows the interaction between the socioeconomic and water resources systems, and the latter is convenient for management and decision-making. In this study, we developed a quantitative method for determining WRCC based on these two theories, and synchronously modified it to transform it into a distributed model that considers spatial differences of the economic development level and water resources conditions of the study area. The research framework was as follows:

(1) The study area was divided into 60 subregions based on the intersections of 18 administrative divisions of Henan Province and 21 third-level sub-catchments of the four first-level river basins. The division of sub-catchments was undertaken according to *The Technical Outline of National Comprehensive Planning of Water Resources* (GIWRHPD, 2006). The reason we divided the study area in this way is that spatial differences among available water resources are determined by the third-level sub-catchments in which rainfall and runoff conditions are different, whereas those of water demand are determined by the administrative divisions and associated with economic development levels. The subregions were then connected by a hydraulic relationship based on the river systems of the study area (Fig. 1), demonstrating the transfer processes of water and contaminants between the subregions.

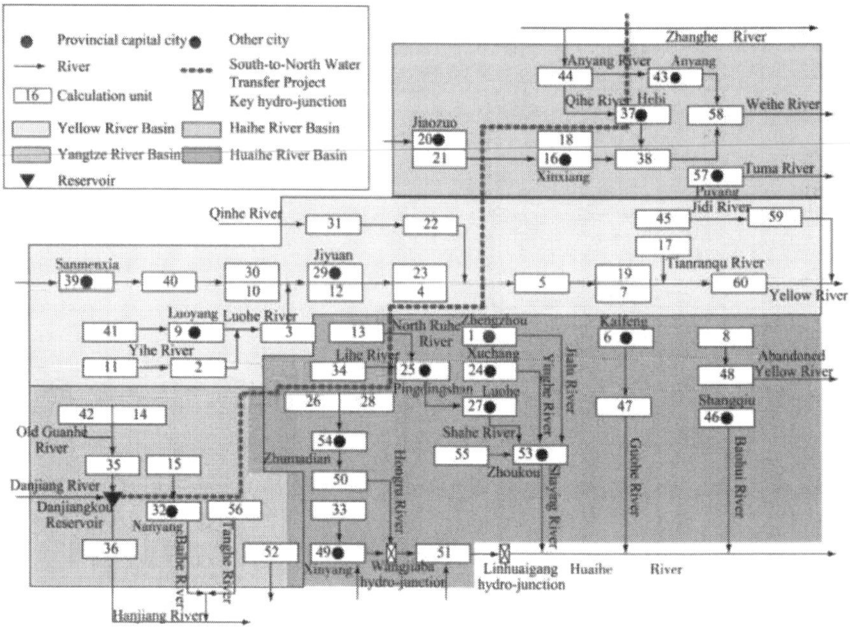

Figure 1. Sketch of water resources calculation nodes.

(2) A distributed hydro-economic model, used to represent the interaction between socioeconomic development and water resources exploitation, was developed based on the large-scale system theory.

(3) Based on the hydro-economic model, an optimal model for calculating WRCC was developed. The model seeks to sustain the maximum socioeconomic scale as the optimization objective, maintain ideal water quality of water function areas, and meet basic water supply and livelihood guarantees as the constraint conditions.

(4) An assessment method with a standard water resources carrying level, which reflected the development potential of the subregions in the future, was developed.

(5) The WRCC of 60 subregions in Henan Province was calculated for different development periods, and the main factors restricting socioeconomic development of Henan Province were identified by these models.

Development of Hydro-economic Model

A hydro-economic model was developed to describe the interaction between the socioeconomic and water resources systems in Henan Province. This model consists of three calculation modules: the socioeconomic system module, water quantity module, and water quality module.

Socioeconomic System Module

This module was developed to simulate the pressure of socioeconomic development on the water resources system inside each subregion, such as socioeconomic development in the future, natural resources consumption, and environmental pollution. This module consists of three components:

(1) Forecast of socioeconomic development level: This was used to forecast the future socioeconomic scale according to the planning development goals of Henan Province. To describe the hydro-economic interaction, it is essential to select several representative socioeconomic indices to reflect the intensity of human activity and its influence on the environment. In this study, the following indices were selected: (a) indices A: population indices, such as total population, the urbanization ratio, the

growth rate of the population, and the floating population; (b) indices B: economic indices, such as GDP, the proportion of the economy made up by the three industries, the increasing rate of GDP, livestock number, and grain yield; (c) indices C: indices that were used to reflect the levels of resource consumption and pollution discharge, such as interbasin water transfer, water use quota, water consumption rate, and the pollutant discharge coefficient of various industries; and (d) indices D: indices that were used to reflect the development level of technology, such as the wastewater treatment efficiency, the wastewater reuse rate, the industrial water recycling rate, and the irrigation water utilization coefficient. In this study, we selected 2010 as the baseline year and 2011 – 2030 as the planning years. Indices A and B were used to forecast the socioeconomic scale of the study area in the planning years, and indices C and D were used to simulate the conditions of water resources supply and use as well as pollutant discharge. In the forecast of the socioeconomic scale, the indices of socioeconomic development level (e.g., the urbanization ratio, the growth rate of the population, and the proportion of the economy made up by the three industries) of the study area were first analyzed and forecasted on the basis of the statistical data from 1980 to 2000. Then, the values of the socioeconomic scale indices (e.g., total population, GDP, livestock number, and grain yield) were calculated according to the functional relationship between them and the parameters above. Finally, the forecasted results were verified and adjusted according to the target values.

(2) Calculation of water use (or demand) and consumption: In the baseline year, the productive (including industrial, agricultural, and service industrial), domestic (including urban domestic and rural domestic), ecological, and total water uses were obtained from the statistical data of water conservancy departments; in the planning years, the productive, domestic, ecological, and total water demands were forecasted with the water quota method, which constructs a functional relationship between the socioeconomic or ecological indices and water demand quota. The corresponding water consumption was then calculated by multiplying the water use (or demand) by the water consumption rate.

(3) Calculation of pollutant discharge: The pollutant discharges arising from point sources (including industrial and urban domestic pollutant discharges) and non-point sources were calculated with the pollutant-discharging coefficient method, which constructs a functional relationship between the socioeconomic indices and pollutant-discharging coefficients. The determination of the pollutant-discharging coefficients was based on comprehensive consideration on the industrial source, domestic lifestyle,

and pollution source treatment level. The river pollution load was then calculated by multiplying pollutant discharge quantities into rivers by the river load coefficients derived empirically on the basis of the degradable characteristic of pollutants, distance from the pollution source to rivers, and canalization level of drainage channels.

Water Quantity Module

First, the hydraulic connection among the subregions was established, and a sketch of the water resources calculation node was made (Fig. 1). All of the variables pertaining to water resources in a subregion, such as the volume of inflow and outflow, self-produced and transferred water, and water consumption, were determined during a certain period of time. According to the relationships between the unknown variables and the given variables, the unknown variables were expressed by the approximate functions of the given variables and then input into the model. The unknown parameters were identified and inserted into the model. According to the calculated multiple correlation coefficients, the regression results were evaluated, and the identified parameters were input into the model. The values of all the water volumes were calculated and checked to determine whether the calculations conformed to the law of mass conservation. In this study, the precipitation and inflow data of 33 hydrological stations in Henan Province were used in the calculation of water quantity. For any time period Δt, there is a water balance equation as follows:

$$P+QT+Q_{in}=E+Ei+Ea+E_{ud}+E_{rd}+Q_{out}+\Delta V \tag{1}$$

where P is the volume of precipitation, in m^3; Q_T is the volume of transferred water, in m^3; Q_{in} is the volume of water inflow, in m^3; E is the volume of evapotranspiration (including water surface and land surface evaporations), in m^3; E_i, E_a, E_{ud}, and E_{rd} are the volumes of industrial, agricultural, urban domestic, and rural domestic water consumptions, in m^3; Q_{out} is the volume of water outflow, in m^3; and ΔV is the volume change in water storage, in m^3.

Water Quality Module

Along with the water quantity module, a module for water quality was also developed based on the law of mass conservation. Because organic pollution was the most outstanding problem of surface water pollution in Henan Province, the permanganate index (COD_{Mn}) was selected as the only water quality index in the following calculation process. In this study, the COD_{Mn} concentrations at 21 monitoring sections in Henan Province were used in the calculation of water quality. For any time period Δt, the following pollutant balance equation was satisfied:

$$\Delta CVs = (1-\beta)(Q_{si}C_{in}+S_{ps}+S_{ns})-(Q_{so}+Wu)C_{out} \tag{2}$$

where Q_{si} and Q_{so} are inflow and outflow volumes of surface water, respectively, in m^3;C_{in} and C_{out} are COD_{Mn} concentrations of inflow and outflow, respectively, in mg/m^3; ΔC is the change in COD_{Mn} concentrations over the time period Δt, in mg/m^3; S_{ps} andS_{ns} are river pollutant loads from point sources and non-point sources, respectively, in mg; V_s is the volume of surface water storage, in m^3; W_u is the volume of total water use, in m^3; and β is the comprehensive reduction rate of COD_{Mn}, which is related with river length, flow velocity, and pollutant characteristics, and determined by a one-dimensional steady-state pollutant migration and transformation equation, which is dimensionless (Dou et al., 2010).

$$\beta = f(x,u,k) = \frac{Q_{si}C_{in}\left(1-e^{-\frac{kx_1}{u}}\right)+\left(S_{ps}+S_{ns}\right)\left(1-e^{-\frac{kx_2}{u}}\right)}{Q_{si}C_{in}+S_{ps}+S_{ns}} \tag{3}$$

where x_1 is the river length from the inflow section to the outflow section of a subregion, in m; x_2 is the river length from the sewage outfall section to the outflow section, in m; u is the average flow velocity, in m/s; and k is the degradation coefficient of COD_{Mn}, in d^{-1}.

Coupling of Modules

The modules of the water quantity, water quality, and socioeconomic system are correlated and interdependent, and they must be coupled to simulate and forecast the interaction between the socioeconomic and water resources systems. The coupling relationship of the modules is illustrated in Fig. 2. In the analysis of the entire study area, based on the input–output relationship of each subregion shown in Fig. 1, all subregions were combined.

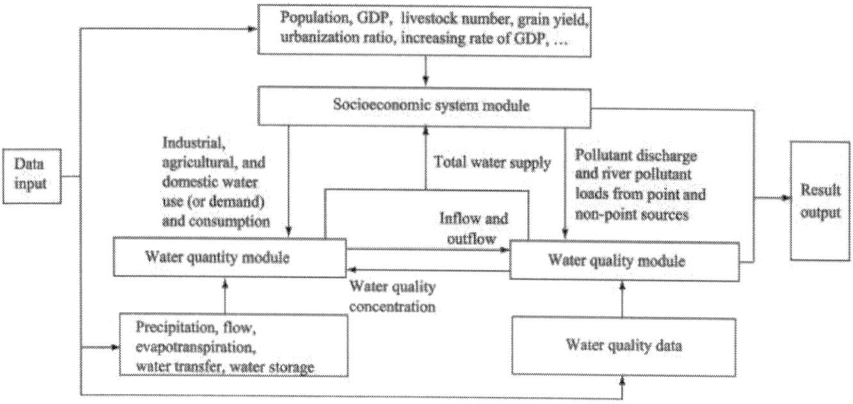

Figure 2. Coupling of modules.

Development of WRCC Quantification Model

The objective of the WRCC quantification model is to find the maximum socioeconomic scale of Henan Province. The socioeconomic scale is usually represented by a series of indices, i.e., urban population, rural population, GDP, industrial added value, agricultural added value, and grain yield. If these indices are included in the objective function, solution of the model will become one solution of the multi-objective optimization problems. For the sake of a convenient solution, all of these indices change proportionally with one another during the search for optimal values. Therefore, the objective function is simplified as a single-objective function based on a change ratio α and can be expressed as follows:

$$Q_{WRCC} = \max f(\alpha(Ppop, RGDP, \cdots)) \tag{4}$$

where Q_{WRCC} is the value of WRCC; α is the change ratio, which is a percentage used to minimize or magnify the socioeconomic scale when solving the WRCC quantifying model; P_{pop} is the population; and R_{GDP} is the value of GDP, in RMB. The constraints of the WRCC quantifying model are as follows.

(1) Simulation of hydro-economic interaction: The hydro-economic model described above is embedded in the WRCC quantifying model as an important constraint. It plays an important part in constructing the relationship between the objective function and other constraints, realizing the transmission and feedback of data between the socioeconomic and water resources systems.

(2) Water supply constraint: Total water use of each subregion must be less than or equal to its available water supply.

$$Wp + We + Wd \leq Was \quad (5)$$

where W_p, W_e, and W_d are, respectively, the volumes of productive, ecological, and domestic water use, in m^3, and W_{as} is the volume of available water supply, including available surface water, available groundwater, transferred water from outside subregions, and wastewater reuse, in m^3.

(3) Water quality concentration constraint: The calculated value of COD_{Mn} concentration in representative water function areas in each subregion must be less than or equal to the corresponding control objective value:

$$C \leq C_S \quad (6)$$

where C is the calculated value of COD_{Mn} concentration, and C_S is the control objective value of COD_{Mn} concentration, in mg/L. In this study, there were 60 water function areas selected as the representative concentration control nodes, according to the following principles: (a) these water function areas were the major stream segments in the subregions; (b) these water function areas received the greatest mass of flow and pollutant afflux in the subregions where they were located; and (c) for subregions in which hydraulic connections were very complex, the number of representative water function areas was increased appropriately.

(4) Socioeconomic development level constraint: Per capita GDP and per capita share of grain in each subregion must be greater than or equal to a certain living standard:

$$Ra \leq R_S \quad (7)$$

$$F_a \leq F_S \quad (8)$$

where R_a and R_S are the calculated values and the minimum living standards of per capita GDP, respectively, in RMB; and F_a and F_S are the calculated values and the minimum living standards of per capita share of grain, respectively, in tons. The minimum living standards were obtained from *The Future Development Outline of Henan Province* (PGHP, 2010).

The WRCC quantifying model is composed of the objective function and constraints. Applying the numerical iteration method (NIM), an approximate optimal solution that meets all of the constraints is obtained. When searching for the optimum value with the NIM, the actual (or forecasted) socioeconomic data input as the initial value is enlarged or reduced by adjusting α in a certain proportion and judging whether the result can satisfy the constraints. The search process is repeated until the difference in the α value of two adjacent searches is less than a certain error coefficient (the value is 0.01 in this study). At this time, the adjusted socioeconomic scale is the WRCC.

When solving the WRCC quantifying model, there arises a situation in which the Q_{WRCC} of a downstream subregion is zero or very small due to the severe pollution of inflow from upstream. The reason is that, because all the subregions in the study area are closely linked through hydraulic relationships, water use and pollutant discharge of a subregion will have a certain impact on downstream subregions, especially those in urban districts. Hence, for downstream subregions, this further diminishes the socioeconomic scale of adjacent upstream subregions, in addition to allowing them to adjust their own scale. Finally, a balance point, which takes into account the coordinated development between the upstream and downstream subregions, may be obtained.

Assessment of Water Resources Carrying Level

The water resources carrying level is used to express the degree of socioeconomic development pressure on water resources systems (Dou et al., 2010). The water resources carrying level is the ratio of the actual (or forecasted) socioeconomic scale in the baseline (or planning) year to the WRCC. When the water resources carrying level is greater than 1.0, the actual (or forecasted) socioeconomic scale has exceeded the WRCC, and the overload becomes more severe as it increases; when it is equal to 1.0, the socioeconomic scale is at the threshold value of WRCC; and when it is less than 1.0, the socioeconomic scale is within the WRCC range, and the development potential becomes greater as it decreases. Furthermore, the water resources carrying level is divided into five categories: when it is less than or equal to 0.6, it falls into the fully loaded category; between 0.6 and 1.0, it falls into the suitably loaded category; between 1.0 and 1.5, it falls into the mildly overloaded category; between 1.5 and 2.0, it falls into the moderately overloaded category; and when it is greater than or equal to 2.0, it falls into the severely overloaded category.

RESULTS AND DISCUSSION

Calibration and Validation

In the simulation of the hydro-economic interaction process, the twenty sensitive parameters listed in Table 1 were selected for Henan Province's hydro-economic model, including seven socioeconomic parameters, seven water quantity parameters, and six water quality parameters.

Table 1. Calibrated parameters of hydro-economic model in Henan Province.

Category	Symbol	Unit	Definition	Range of value
Socioeconomy	a_u	m^3/(capita·year)	Urban domestic water use quota	90–145
	a_r	m^3/(capita·year)	Rural domestic water use quota	60–90
	a_i	$m^3/(10^4$RMB·year)	Industrial water use quota	20–86
	a_t	$m^3/(hm^2$·year)	Agricultural water use quota	2 100–5 850
	G_y	kg/(hm^2·year)	Grain output per hectare	5 500–11 500
	R_{ru}	%	Wastewater reuse rate	0–40
	T_{ws}	%	Discharging rate of wastewater whose concentration meets corresponding standard after treatment	60–100
Water	R_P	–	Runoff generation coefficient	0.28–0.50

quantity	I_{ps}	–	Infiltration recharge coefficient of precipitation and surface water	0.10–0.30
	A_{wr}	–	Comprehensive coefficient of available water resources	0.40–0.65
	h_u	%	Urban domestic water consumption rate	20–35
	h_r	%	Rural domestic water consumption rate	85–100
	h_i	%	Industrial water consumption rate	20–40
	h_a	%	Agricultural water consumption rate	70–92
Water quality	d_i	kg/(10^4RMB·year)	Industrial pollutant discharge coefficient	7.0–12.5
	d_u	kg/(capita·year)	Urban domestic pollutant discharge coefficient	2.0–10.3
	P_{np}	mg/L	Pollutant producing coefficient of non-point source	14.0–24.0
	λ_{ns}	–	River load ratio of non-point source	0.45–0.60
	λ_{ps}	–	River load ratio of point source	0.07–0.15
	k	d^{-1}	Degradation coefficient of COD_{Mn}	0.05–0.20

In order to demonstrate the simulation effect of the hydro-economic model, three indices, average relative error, the correlation coefficient, and the efficiency coefficient (Zhang et al., 2013), were used to measure the simulation precision. Both discharge and COD_{Mn}concentration were simulated using this model, and the applicability of the model was evaluated by investigating the processes of change in the variables. The simulation results are summarized in Table 1. In the water quantity simulation, there were 21 stations whose absolute value of average relative error values were less than 15%, accounting for 64% of all the stations. The average correlation coefficient was 0.72, and the average coefficient of efficiency was 0.45. In the water quality simulation, there were 13 sections, whose absolute value of average relative error values were less than 45%, accounting for 62% of all the sections, and 15 sections whose correlation coefficients were greater than 0.40, accounting for 71% of all the sections. According to the results of the simulation, we concluded that the model was a satisfactory and reasonably representative of the hydro-economic interaction.

WRCC in Baseline Year

We selected 2010 as the baseline year, when the WRCC of 60 subregions in Henan Province under actual conditions was determined using the WRCC quantification model. The calculation conditions applied in the WRCC quantification model in the baseline year were as follows: (a) hydrological conditions were adopted based on statistical precipitation, evaporation, and runoff data from 2010; (b) socioeconomic conditions were adopted based on statistical socioeconomic indices, such as population, GDP, livestock number, and grain yield, from 2010; (c) water resources conditions were adopted based on statistical data on water resources exploitation and utilization from 2010; and (d) water environmental conditions were adopted based on monitored water quality concentrations of the rivers and statistical pollutant discharge and pollutant load amounts of rivers from 2010.

Based on these calculation conditions, the WRCC of each subregion was calculated using the model, and the water resources carrying level was determined. Then, the WRCC of 18 cities and of the entire province were aggregated according to the calculation results. The results are shown in Table 2, and the spatial distribution of the water resources carrying level is shown in Fig. 3.

Table 2. WRCC in baseline year under actual conditions.

City	Water resources carrying level	Urban population (10^4capita)	Rural population (10^4capita)	GDP (10^9RMB)	Industrial added value (10^9RMB)	Agricultural added value (10^9RMB)	Grain yield (10^4 t)
Zhengzhou	1.32	363.1	203.4	308.889	152.466	9.083	121.6
Kaifeng	1.15	150.5	274	80.054	31.784	18.915	220.4
Luoyang	0.57	501.5	670.9	391.673	208.764	33.517	421.4
Xinxiang	1.1	218.7	295.1	111.225	56.75	13.775	334.1
Jiaozuo	2.4	72.9	78.9	50.755	32.783	4.072	80.1
Xuchang	1.26	133.6	208.9	104.677	67.387	11.933	218.8
Pingdingshan	0.56	369.9	542.1	237.732	148.87	20.62	354.5
Luohe	1.61	62.1	99	42.363	28.147	5.393	104.5
Jiyuan	1.49	24.6	31.7	27.302	19.446	1.123	15.3
Nanyang	0.46	769.1	1 559.8	421.337	196.322	86.624	1 261.0
Hebi	1.74	31	53.6	21.921	14.141	2.933	67.1
Sanmenxia	0.85	112.9	153.3	95.834	59.756	8.058	72.7
Anyang	2.65	73.8	132.2	46.449	25.346	6.248	131.3
Shangqiu	1.78	132.9	329.9	62.235	25.031	16.714	334.1
Xinyang	0.73	362.8	738.8	148.47	50.84	40.05	799.8
Zhoukou	1.61	196.5	467.3	75.802	30.33	22.666	448.6
Zhumadian	1.63	145	371.3	60.664	22.308	17.286	398.6
Puyang	1.1	92.3	241.6	64.524	39.598	8.925	208.2
Entire province	0.98	3 813.2	6 451.8	2 351.906	1 210.069	327.935	4 331.1

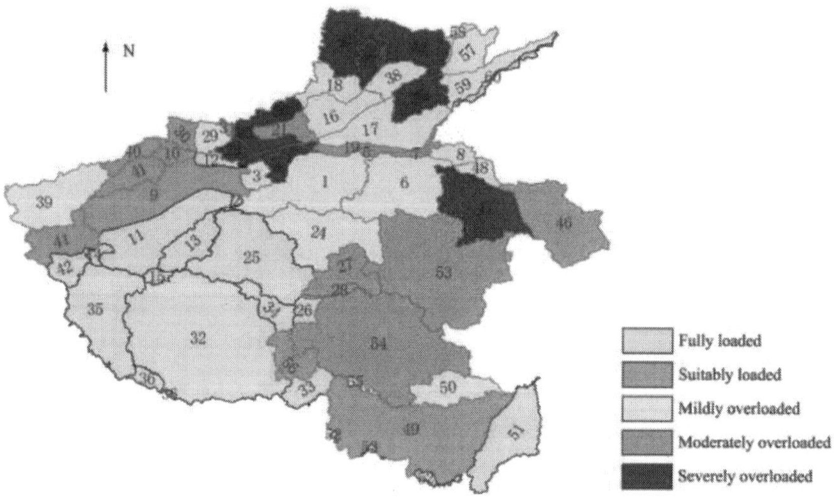

Figure 3. Assessment of water resources carrying level in baseline year.

The results in Table 2 indicate that the water resources carrying level of Henan Province was 0.98 in 2010, which falls into the suitably loaded category. The loaded total population was 10 265.0 × 10^4 and the loaded GDP value was 23 519.06 × 10^8 RMB. These were 102.5% of the actual values of the province (the actual total population was 10 018.0 × 10^4 and the actual GDP value was 22 953.26 × 10^8 RMB in 2010). There were 33 overloaded subregions, accounting for 55% of the total subregions. Of these 33 overloaded subregions, nine subregions fell into the severely overloaded category, seven subregions fell into the moderately overloaded category, and 17 subregions fell into the mildly overloaded category. Despite the fact that 2010 was a wet year and water resources in that year were correspondingly abundant, the actual socioeconomic scale of some subregions was nonetheless found to exceed their WRCC.

The subregions with higher water resources carrying levels included some subregions in the headwaters of the Huaihe River and its major branches, such as subregion 13 (Luoyang City), subregion 32 (Nanyang City), subregion 35 (Nanyang City), subregion 49 (Xinyang City), and subregion 51 (Xinyang City). These subregions lie in the southern part of Henan Province, where there is abundant precipitation as well as low water consumption and pollutant discharge, and the high water resources carrying level of these subregions was due to their slower production. The others were the subregions along the main stream of the Yellow River, such as subregion 2 (Zhengzhou City), subregion 7 (Kaifeng City), subregion 9 (Luoyang City), subregion 19 (Xinxiang City), and subregion 60 (Puyang City). These subregions have abundant water flowing through them even though they generate less water because of scarce precipitation and small catchment areas. In the meantime, water consumption and pollutant discharge in these subregions are low due to their small socioeconomic scale.

The subregions with lower water resources carrying levels included some subregions in the Haihe River Basin, such as subregion 4 (Zhengzhou City), subregion 20 (Jiaozuo City), subregion 30 (Jiyuan City), subregion 43 (Anyang City), subregion 44 (Anyang City), subregion 45 (Anyang City), subregion 57 (Puyang City), and subregion 59 (Puyang City). The Haihe River Basin is one of the areas that suffers from severe water shortage, in which precipitation and generated water amount are both low. At the same time, these subregions are relatively isolated, and replenishment of water and transferred water both remain at low levels. The others are the subregions in the central urban areas of Henan Province, such as subregion 1 (Zhengzhou City), subregion 6 (Kaifeng City), subregion 17 (Xinxiang City), subregion 24 (Xuchang City), subregion 53 (Zhoukou City), subregion 54 (Zhumadian City), subregion 46 (Shangqiu City), and subregion 47 (Shangqiu City). Although these subregions have certain available water resources, their water consumption and pollutant discharge are large because of their large socioeconomic scale and unreasonable industrial structure. For example, the dominant industries are energy, machinery, and raw material industries in subregion 1 (Zhengzhou City); modern industry in subregion 17 (Xinxiang City); and machinery manufacturing and fur processing industries in subregion 24 (Xuchang City).

WRCC in Planning Years

The planning years were from 2011 to 2030, and 2030 was selected as a key time node to be introduced and compared with the baseline year in this study. The calculation conditions in the planning years were as follows: (a) hydrological conditions based on statistical data on precipitation, evaporation, and runoff at the hydrological frequency of 50%; (b) socioeconomic conditions based on forecasted socioeconomic index values from 2011 to 2030, according to future development goals from the *Central Plains Economic Zone Development Planning* (PGHP, 2012); (c) water resources conditions based on the planning water resources utilization level from 2011 to 2030 according to the *Henan Province Water Resources Integrated Planning* (HWPECC, 2008); and (d) water environmental conditions based on the controlled water quality concentration in rivers and forecasts of pollutant discharge and load amounts of rivers from 2011 to 2030, according to environmental protection goals from the *Henan Province Environmental Protection Planning* (HPEPB, 2012).

Based on the calculation conditions described above, the WRCC of each subregion was calculated and the water resources carrying level was determined. The aggregated results for the entire province from 2011 to 2030 are shown in Table 3, and spatial distribution of the water resources carrying level in 2030 is shown in Fig. 4.

Table 3. Henan Province's WRCC in planning years under forecasted conditions.

Year	Water resources carrying level	Urban population (10^4 capita)	Rural population (10^4 capita)	GDP (10^9RMB)	Industrial added value (10^9 RMB)	Agricultural added value (10^9 RMB)	Grain yield (10^4 t)
2011	1.33	3 000.9	4 727.0	1 912.015	973.744	247.59	4 453.2
2012	1.35	3 057.1	4 489.2	2 118.340	1 068.913	255.47	4 596.9
2013	1.44	3 003.2	4 115.9	2 334.256	1 168.043	262.84	4 611.7
2014	1.47	3 077.3	3 939.9	2 542.576	1 262.611	267.9	4 722.0
2015	1.53	3 087.8	3 696.0	2 723.178	1 342.880	269.03	4 936.2
2016	1.58	3 118.1	3 491.5	2 853.070	1 397.930	264.75	5 016.3
2017	1.46	3 513.9	3 682.7	2 906.453	1 415.681	253.72	4 813.5
2018	1.44	3 705.3	3 635.7	2 914.666	1 411.929	239.7	4 853.0
2019	1.41	3 930.9	3 611.6	2 935.544	1 414.837	227.71	4 857.9
2020	1.36	4 324.0	3 720.2	2 949.926	1 415.066	216.09	4 739.2
2021	1.34	4 392.4	3 633.1	3 005.454	1 418.295	208.14	4 679.8
2022	1.36	4 426.7	3 521.4	3 170.354	1 473.567	207.98	4 705.7
2023	1.35	4 558.9	3 489.0	3 337.749	1 529.610	207.8	4 826.4
2024	1.35	4 658.0	3 430.8	3 533.245	1 598.007	209.09	4 822.4
2025	1.33	4 828.5	3 423.6	3 600.032	1 608.278	202.81	4 865.3
2026	1.32	4 966.0	3 390.4	3 716.260	1 641.145	199.57	4 883.6
2027	1.31	5 105.4	3 357.0	3 744.302	1 635.699	191.93	4 901.1
2028	1.27	5 370.7	3 401.8	3 823.869	1 653.504	187.3	4 994.2
2029	1.3	5 348.6	3 264.0	4 085.163	1 749.585	191.42	5 072.0
2030	1.31	5 408.6	3 180.5	4 131.221	1 753.317	185.37	5 365.1

Note: The water resources carrying level is the average value throughout Henan Province.

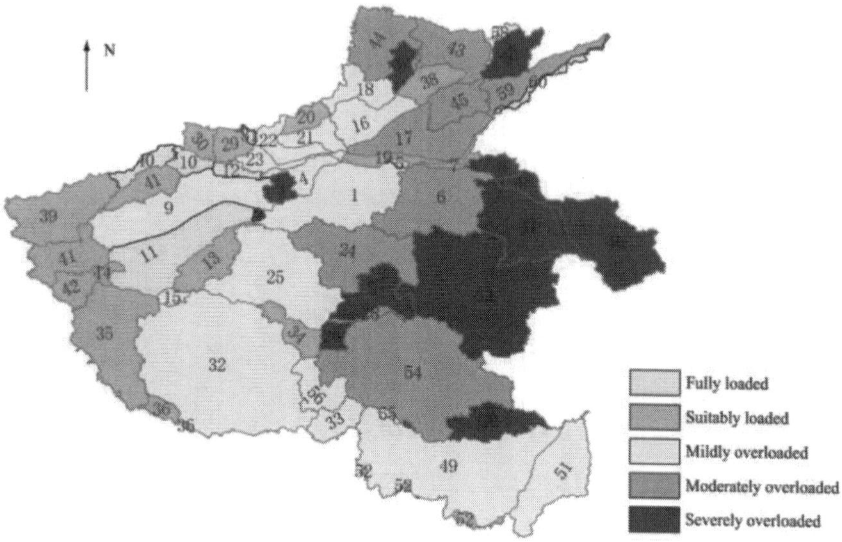

Figure 4. Assessment of water resources carrying level in 2030.

The results in Table 3 indicate that the water resources carrying level of Henan Province will be 1.31 in 2030, which falls into the mildly overloaded category. The loaded total population of the province in 2030 will be 8 589.1 × 10^4, which is a decrease of 1 675.9 × 10^4 from 2010 with a decrease rate of 16.33%. The loaded urban population was predicted to increase by 1 595.4 × 10^4 over that in 2010, with an increase rate of 41.84%. The loaded rural population was predicted to decrease by 3 271.3 × 10^4 from that in 2010, with a decrease rate of 50.70%. This is due to the rapid urbanization ratio predicted to occur from 2010 to 2030, which will result in significant migration from the countryside to the city. The loaded GDP value of the province was predicted to be 41 312.21 × 10^8 RMB in 2030, which is an increase of 17 793.15 × 10^8 RMB over that in 2010, with an increase rate of 75.65% from 2010 to 2030. The proportions of agricultural, industrial, and service industrial added values were predicted to change dramatically, from 13.9: 51.5: 34.6, in 2010 to 4.5: 42.4: 53.1, in 2030. In the future, the service industry will be the dominant industry rather than traditional industry.

There are 38 subregions predicted to be overloaded, accounting for 63% of total number of subregions. Of these 38 subregions, 12 subregions were predicted to be severely overloaded, nine subregions were moderately overloaded, and 17 subregions were mildly overloaded. For 29 subregions, the water resources carrying level was predicted to decrease from 2010 to

2030. However, there were also 16 subregions in which the water resources carrying level was predicted to improve from 2010 to 2030. For example, subregion 20 (Jiaozuo City), subregion 29 (Jiyuan City), subregion 30 (Jiyuan City), subregion 38 (Hebi City), and subregion 39 (Sanmenxia City) will move into the suitably loaded category.

The water resources condition (i.e., available water resources) is a crucial factor restricting socioeconomic development of water-deficient areas, such as the northern and central areas of Henan Province. According to the calculation results, both the water resources amount and the WRCC value in 2010 are greater than those predicted in 2030. The main reason is that the hydrological conditions for 2010 are based on actual data, and 2010 was a wet year, while the hydrological conditions for 2030 are based on designed data, and 2030 is predicted to be a normal year. In addition, the influences of future socioeconomic development and technological progress on the WRCC are significant in the following ways: (a) The increase of the urbanization ratio and GDP will lead to the growth of water demand quota and total water demand (Fig. 5(a) and (c)), which will contribute to a decrease in the WRCC. (b) The irrigation water utilization coefficient, industrial water recycling rate, and wastewater reuse rate will improve with the progress of science and technology (Fig. 5(d)), which will contribute to an increase in the WRCC. (c) Total water supply and interbasin water transfer will increase due to the construction of various water transfer projects (Fig. 5(b)), which will contribute to an increase in the WRCC. Because the indices of the increasing rate of GDP, urbanization ratio, irrigation water utilization coefficient, industrial water recycling rate, wastewater reuse rate, and interbasin water transfer will vary temporally and spatially, and the trends in the WRCC vary among the subregions and over time. From the temporal perspective, the number of overloaded subregions exhibits an increasing trend over the period from 2011 to 2016, and a decreasing trend over the period from 2017 to 2030 (Fig. 6(a)), and the loaded population exhibits an opposite trend, decreasing at first and then increasing (Fig. 6(b)). In the former period, the negative factors (e.g., the increasing rate of GDP and urbanization ratio) are more influential than the positive factors (e.g., the irrigation water utilization coefficient, industrial water recycling rate, wastewater reuse rate, and interbasin water transfer) when the water resources conditions are consistent, but in the later period, the positive factors are more influential on the WRCC than the negative factors. A significant reason is that the middle route of the South-to-North Water Transfer Project of China will be completed and begin transferring water in 2017. With the construction of this project, Henan Province could increase additional freshwater by 3.769×10^9 m^3 each year, accounting for 14.2% of the province's available water supply in 2017. Basically, the water demand of 11 large and medium-sized cities in Henan Province along the middle route will be met, but the water demand of other cities will remain the same. Meanwhile, the loaded GDP value exhibits a consistently increasing trend over the period from 2011 to 2030 (Fig. 6(c)). On the whole, the indices of socioeconomic development and urbanization show increasing trends, but those of the rural population and the proportion of agricultural added value show decreasing trends (Fig. 6(d)).

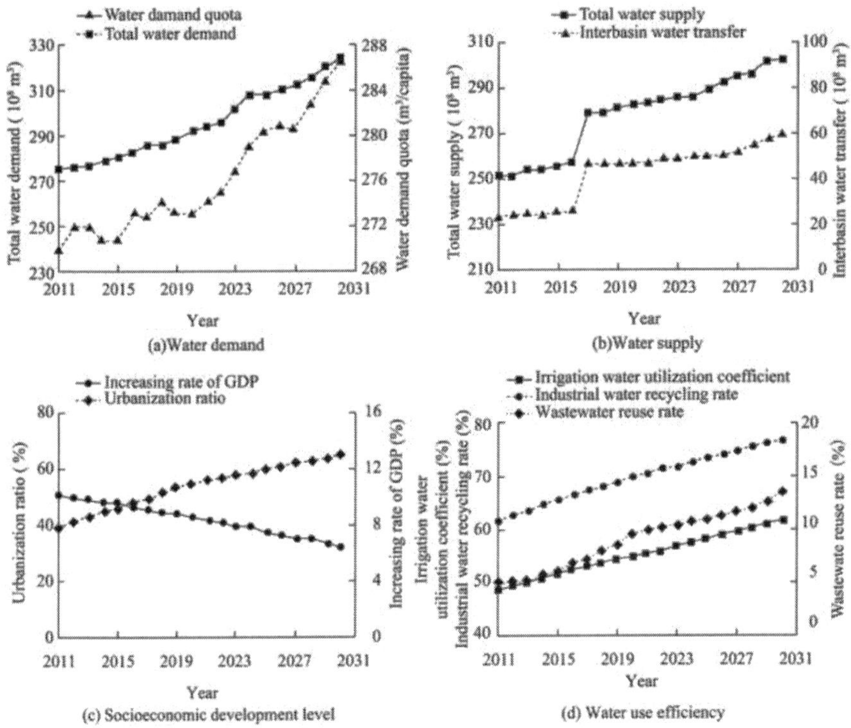

Figure 5. Trends of socioeconomic development and water resources indices in planning years.

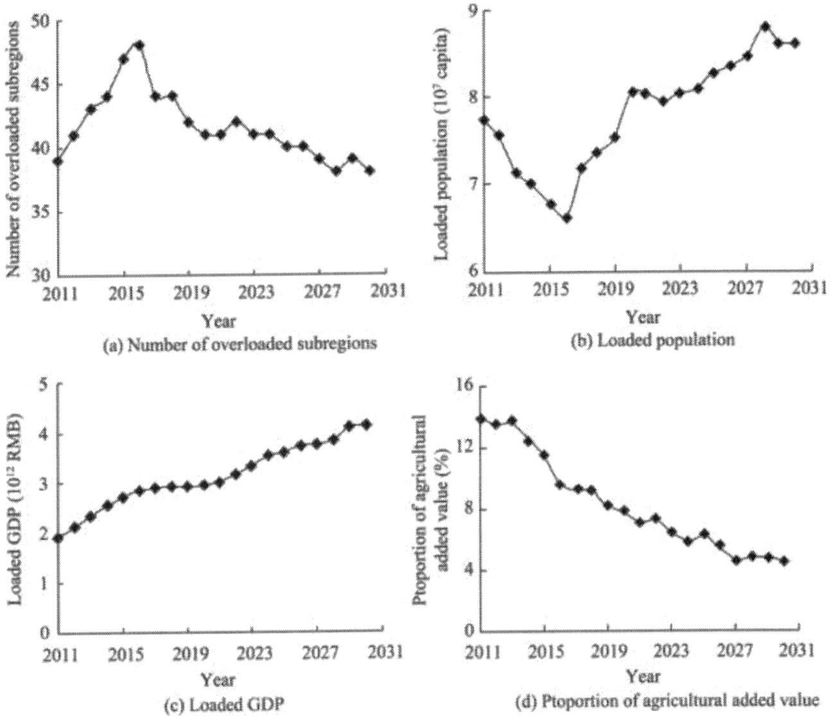

Figure 6. Trends of socioeconomic indices and number of overloaded subregions in planning years.

CONCLUSIONS

In this paper, the concept of WRCC was discussed, and some calculation and evaluation methods for WRCC were presented. Using these methods, the WRCC of Henan Province was determined for the baseline and planning years, and the following conclusions are drawn:

(1) A distributed quantitative model for WRCC based on the large-scale system theory and optimization method was developed, and Henan Province's WRCC was calculated for different development periods.

(2) According to the simulation results of the model, there were 33 overloaded subregions in 2010; the loaded total population was $10\,265.0 \times 10^4$, and loaded GDP value was $23\,519.06 \times 10^8$ RMB. The province's water resources carrying level is suitably loaded.

(3) Based on the planning development goals, the WRCC from 2011 to 2030 was calculated. According to the simulation results, there will be 38 subregions overloaded in 2030, the loaded total population will be 8 589.1 × 10^4, and the loaded GDP value will be 41 312.221 × 10^8 RMB. The province's water resources carrying level will be mildly overloaded.

(4) WRCC is influenced by many factors, of which available water resources is the most crucial. The main factors affecting the WRCC are different from 2011 to 2030. In the period from 2011 to 2016, the negative factors (e.g., the increasing rate of GDP and urbanization ratio) that contribute to a decrease of the WRCC are more influential than the positive factors (e.g., the irrigation water utilization coefficient, industrial water recycling rate, wastewater reuse rate, and interbasin water transfer) that contribute to an increase of the WRCC when the water resources conditions are consistent, and the converse is true for the latter period from 2017 to 2030.

REFERENCES

1. Clarke, A.L., 2002. Assessing the carrying capacity of the Florida Keys. Popul. Environ. 23(4), 405e418. http://dx.doi.org/10.1023/A:1014576803251.
2. Dang, X.H., Guo, B.L., 2012. Emergy measures of carrying capacity and sustainability of a target region for an ecological restoration programme: a case study in Loess Hilly Region, China. J. Environ. Manag. 102, 55e64. http://dx.doi.org/10.1016/j.jenvman.2012.02.023.
3. Dou, M., Zuo, Q.T., Hu, R., 2010. Carrying capacity of water environment in the Huaihe River basin. Adv. Water Sci. 21(2), 248e254 (in Chinese). Falkenmark, M., Lundqvist, J., 1998. Towards water security: political determination and human adaptation crucial. Nat. Resour. Forum 21(1), 37e51. http://dx.doi.org/10.1111/j.1477-8947.1998.tb00708.x.
4. Feng, L.H., Zhang, X.C., Luo, G.Y., 2008. Application of system dynamics in analyzing the carrying capacity of water resources in Yiwu City, China. Math. Comput. Simul. 79(3), 269e278. http://dx.doi.org/10.1016/j.matcom.2007.11.018.
5. Feng, S.Y., Chen, S.J., Huo, Z.L., Li, W.C., 2006. Review on the present situation and future prospect of water resources carrying capacity in China. J. East China Inst. Technol. 29(4), 301e306 (in Chinese).

6. General Institute of Water Resources and Hydropower Planning and Design (GIWRHPD), Ministry of Water Resources, 2006. The Technical Outline of National Comprehensive Planning of Water Resources. GIWRHPD, Beijing.
7. Graymore, M.L.M., Sipe, N.G., Rickson, R.E., 2009. Sustaining human carrying capacity: a tool for regional sustainability assessment. Ecol. Econ. 69(3), 459e468. http://dx.doi.org/10.1016/j.ecolecon.2009.08.016.
8. Henan Province Environmental Protection Bureau (HPEPB), 2012. Henan Province Environmental Protection Planning. HPEPB, Zhengzhou. Henan Water and Power Engineering Consulting Co., LTD. (HWPECC), 2008.
9. Henan ProvinceWaterResources Integrated Planning. HWPECC, Zhengzhou. Hunter, C., 1998. Perception of the sustainable city and implications for fresh water resources management. Int. J. Environ. Pollut. 10(1), 84e103.
10. Lane, M., Dawes, L., Grace, P., 2014. The essential parameters of a resourcebased carrying capacity assessment model: an Australian case study. Ecol. Model. 272, 220e231. http://dx.doi.org/10.1016/j.ecolmodel. 2013.10.006.
11. Li, L.J., Guo, H.C., Chen, B., 2000. Water resources supporting capacity of Chaidamu Basin. Environ. Sci. 21(2), 20e23. http://dx.doi.org/10.13227/ j.hjkx.2000.02.005 (in Chinese).
12. Liu, H.M., 2012. Comprehensive carrying capacity of the urban agglomeration in the Yangtze River Delta, China. Habitat Int. 36(4), 462e470. http:// dx.doi.org/10.1016/j.habitatint.2012.05.003.
13. Liu, R.Z., Borthwick, A.G.L., 2011. Measurement and assessment of carrying capacity of the environment in Ningbo, China. J. Environ. Manag. 92, 2047e2053. http://dx.doi.org/10.1016/j.jenvman.2011.03.033.
14. Long, T.R., Jiang, W.C., He, Q., 2004. Water resources carrying capacity: newperspectives based on eco-economic analysis and sustainable development. J. Hydraulic Eng. 35(1), 38e45 (in Chinese).
15. National Research Council (NRC), 2002. A Review of the Florida Keys Carrying Capacity Study.
16. National Academy Press, Washington, D.C. Ofoezie, I.E., 2002. Human health and sustainable water resources development in Nigeria: schistosomiasis inartificial lakes. Nat. Resour. Forum 26(2), 150e160. http://dx.doi.org/10.1111/1477-8947.00015.
17. Oh, K., Jeong, Y., Lee, D., Lee, W., Choi, J., 2005. Determining development density using the urban carrying capacity assessment system. Landsc. Urban Plan. 73(1), 1e15. http://dx.doi.org/10.1016/j.landurbplan.2004.06.002.
18. The People's Government of Henan Province (PGHP), 2010. The Future Development Outline of Henan Province. PGHP, Zhengzhou.
19. The People's Government of Henan Province (PGHP), 2012. Central Plains Economic Zone Development Planning Outline. PGHP, Zhengzhou.

20. Prato, T., 2009. Fuzzy adaptive management of social and ecological carrying capacities for protected areas. J. Environ. Manag. 90(8), 2551e2557. http://dx.doi.org/10.1016/j.jenvman.2009.01.015. Queensland University of Technology (QUT), 2012. Carrying Capacity Dashboard: a Resource-based Indicator of Regional Self Sufficiency. http://dashboard.carryingcapacity.com.au/ (Retrieved Apr. 13, 2015).
21. Rees, W.E., 1997. Urban ecosystems: the human dimension. Urban Ecosyst. 1(1), 63e75. Seidl, I., Tisdell, C., 1999. Carrying capacity reconsidered: from Malthus' population theory of cultural carrying capacity. Ecol. Econ. 31(3), 395e408. http://dx.doi.org/10.1016/S0921-8009(99)00063-4.
22. Shi, Y.F., Qu, Y.G., 1992. Carrying capacity of water resources in Urumqi River Basin and its rational utilization. Science Press, Beijing, pp. 128e139 (in Chinese). Xia, J., Zhu, Y.Z., 2002.
23. The measurement of water resources security: a study and challenge on water resources carrying capacity. J. Nat. Resour. 17(3), 263e269 (in Chinese).
24. Xu, Z., Cheng, G., 2000. A study on the water resources carrying capacity by using the method of multiobjective optimization model: taking the Heihe River as an example. J. Lanzhou Univ. Nat.(Nat. Sci.) 36(2), 122e132 (in Chinese).
25. Zai, S.M., Wen, J., Wu, F., Sun, H., 2011. Evaluation of water resources carrying capacity in Xinxiang City. J. Hydraulic Eng. 42(7), 783e788 (in Chinese).
26. Zhang, B.C., Guo, F., 2006. Reviews on the research of water resources carrying capacity. Shanghai Econ. Rev. 10(6), 39e43 (in Chinese).
27. Zhang, Y.X., Chen, M., Zhou, W.H., Zhuang, C.W., Ouyang, Z.Y., 2010. Evaluating Beijing's human carrying capacity from the perspective of water resource constraints. J. Environ. Sci. 22(8), 1297e1304. http://dx.doi.org/10.1016/S1001-0742(09)60253-9.
28. Zhang, Y.Y., Xia, J., Shao, Q.X., Zhai, X.Y., 2013. Water quantity and quality simulation by improved SWAT in highly regulated Huai River Basin of China. Stoch. Environ. Res. Risk Assess. 27(1), 11e27. http://dx.doi.org/10.1007/s00477-011-0546-9.

CITATION

Ming Dou, Jun-xia Ma, Gui-qiu Li, Qi-ting Zuo, Measurement and assessment of water resources carrying capacity in Henan Province, China, Water Science and Engineering, Volume 8, Issue 2, April 2015, Pages 102-113, ISSN 1674-2370, http://dx.doi.org/10.1016/j.wse.2015.04.007.

CHAPTER 6

Climate-Change Impacts on Water Resources and Hydropower Potential in the Upper Colorado River Basin

M. Kopytkovskiy[1, 2]*, M. Geza*[2]*, J.E. McCray*[2]

[1]Civil and Environmental Engineering, Colorado School of Mines, 1500 Illinois St., Golden, CO 80401, USA

[2] Parker Water and Sanitation District, 18100 E. Woodman Drive, Parker, CO 80134, USA

ABSTRACT

Study Region

The Upper Colorado River Basin (UCRB), comprised of the Colorado and Gunnison River basins, is the prime water source for much of the western United States.

Study Focus

Future climate change models were used to drive a hydrologic model of the UCRB to evaluate future water resources and hydropower potential of the basin, using three different climate projections. The Intergovernmental Panel on Climate Change (IPCC) emission scenarios, the A2-business as usual, and the B1-reduced emissions scenarios were evaluated. More than 4500 water diversions and 17 reservoirs were incorporated into the hydrologic model.

New Hydrological Insights for the Region

Precipitation projections from climate models vary up to 16%; flow projections revealed greater differences, up to 50%. The climate models projected increase in temperature at low elevations with extreme seasonality at high elevations, although summer temperatures increased at all elevations. The models projected a 60% decline in precipitation at lower elevations and a 74% increase at high elevations, although precipitation declined during the summer months at all elevations. Using the A2 scenario an overall decrease in annual flow was predicted, attributed to a reduction in precipitation and increasing temperature trends; however, this was not consistent during the winter months, which showed an increase in precipitation at high elevations and a modest temperature increase during the winter and resulted in an increase in stream flow. The responses to climate change on reservoir levels varied basin-wide due to variability in precipitation, evapotranspiration, and stream flow. Simulations indicated that water levels in Blue Mesa Reservoir (the largest reservoir in the UCRB) would decline by more than 70% with increasing annual temperatures. Reservoirs with smaller surface areas to the volume ratio were not significantly impacted by evapotranspiration. Our results indicate that hydropower management strategies in the UCRB must adapt to potential climate change, but the required adaptations are dependent on several factors including reservoir size and location.

INTRODUCTION

Potential impacts of climate change on water resources are usually assessed by applying climate projections (temperature and precipitation) derived from global circulation models (GCM) using a hydrologic model. In this study we investigate the impact of global climate change on water resources and reservoir levels (as related to hydropower potential) on the Upper Colorado River Basin (UCRB) using state-of-the art climate projections and an integrated hydrologic model. The Colorado River is critical to the water resources of the southwestern United States. Already almost wholly allocated (Christensen et al., 2004), water management and supply issues of the basin are projected to be exacerbated by climate change (Nash and Gleick, 1991). Multiple studies have shown that the UCRB will respond to a temperature increase with an increase in the rain-to-snow ratio, increased winter runoff, and earlier spring snow melt. Although precipitation predictions vary, most studies agree in the overall reduction of runoff in the basin (Barnett and Pierce, 2009). A previous study on the basin by Christensen et al. (2004), demonstrated that there will be a 10–30% reduction in stream flow. These changes have already resulted in reduced storage levels of two significant reservoirs downstream, Lake Mead and Lake Powell, and are expected to impact the management of flow and reservoir regimes. Christensen et al. (2004) compared the impact of climate change on the Colorado River Basin using three 105-year future climate scenarios based on a "business-as-usual" (BAU) greenhouse gas (GHG) emission scenario to a static 1995 GHG simulation using the Variable Infiltration Capacity (VIC) model. The study concluded there would be a 0–10% increase in flow in the northwest portion of the river in Arizona. Christensen and Lettenmaier (2007) used an ensemble approach to characterize the hydrologic response to climate change using the Colorado River Reservoir Model (CRRM). They employed 11 GCMs and A2 and B1 climate scenarios. A2 is a business as usual (BAU) scenario that evaluates untamed CO_2 levels (850 parts per billion) until the year 2100, whereas B1 assumes a considerable effort to minimize emissions to achieve a moderate atmospheric carbon dioxide concentration of about 550 parts per billion. Although the authors used 11 GCMs, only 11 diversion points were considered to model the most critical junctures, in contrast to the over 4500 diversions considered in this study. The authors determined that an average increase in temperature of approximately 3–4 degrees Fahrenheit (°F), combined with a decrease in precipitation between 1% and 2%, resulted in a decrease in mean runoff to about 11%. Under the B1 scenario, it was predicted that there will be an average increase in temperature of approximately 2.52 °F, a change in precipitation between −1% and 1%, and a decrease in mean runoff up to 8%. McCabe and Wolock (2007) used a water balance model to describe streamflow changes and their impacts on long-term water sustainability of the Southwest. Only considering the warming effects of a changing climate, they concluded that the Colorado River Basin (CRB) will experience future water supply shortages. The United States Bureau of Reclamation (USBR) recently analyzed the basin under varying future scenarios, emphasizing the impact of evapotranspiration (ET). Using bias-corrected statistically downscaled data (BCSD) and changes in ET rates, they predicted a 6–13% decrease in runoff for the Gunnison River (Miller et al., 2011). Raff et al. (2009) used a statistically downscaled monthly precipitation model, temporally disaggregated into 6-h weather forcing at 1/8° spatial resolutions for flood frequency analysis. They concluded a cumulative increase in future predictions of annual flood due to an increase in extreme events. Raff et al. (2009) also noted a need for more studies to benefit from recent advances in GCM data. Rasmussen et al. (2011) analyzed runoff and snowfall trends over Colorado using higher-resolution models to better simulate "orographic precipitation, snowpack accumulation, ablation, evaporation and runoff processes". These authors used the Weather Research and Forecasting (WRF) weather and climate model for regional analysis with the Pseudo Global Warming (PGW) approach at 2 km grid spacing with SNOTEL observations. Results indicated an increase in snowfall at higher elevations, due to the presence of more moisture, with a 10–15% increase in precipitation at the Colorado River headwaters due to the rain shadow effect.

Our current study analyzes changes in precipitation, temperature, streamflow and reservoir storage using current GCM and downscaling approaches (discussed later) with a watershed and reservoir analysis model. Our work considers both the Colorado River and the Gunnison River basins with contrasting climactic conditions and topography, allowing simultaneous assessment of climate change and climate change impacts across different elevations and regions. To the best of our knowledge, our study is the first to account for over 4500 stream diversions, which provides a more robust model for future water rights assessment. Our study shares some common methods used in prior work; we use multiple GCMs and climate scenarios with a physically based hydrologic model that includes an integrated 2D reservoir model. Yet our work on the UCRB differs from previous studies in that it uses the results from a newly developed downscaling method Bias-Corrected Constructed Analogues (BCCA) from three climate models (MIR, MIROC3.2, and CGCM3.1) under two climate scenarios to drive our hydrologic model, which provides a broader and better insight about the impacts of climate change in stream flow and reservoir levels given the uncertainties in the climate models.

Description of Study Area

The Upper Colorado River basin (UCRB), shown in Fig. 1, includes the Colorado River Basin in the North and the Gunnison River basin in the South. It is a mountainous plateau ranging from 4850 to 13,000 feet in elevation, comprised of valleys, canyons, and mountain ranges. The UCRB has a total drainage area greater than 17,800 square miles with a mean flow of over 3000 cfs. The mean flow is approximately split equally between the Colorado River and the Gunnison River. The Colorado River supplies water to seven states and two nations, and irrigates more than 3 million acres of farmland (Barnett and Pierce, 2009). Most of the supply originates as snowmelt in the headwaters. The river is extensively managed to provide reliable water supply for farming and cities; thus, our model included over 4500 diversions and 17 reservoirs. The UCRB has an arid to semi-arid climate, with more pronounced periods of drought in the last half of the 20th century. Under future climate projections, snow accumulation and precipitation volume are projected to decline, while temperatures are expected to increase, thus increasing regional aridity. It is widely agreed that the UCRB will experience a 2–4 °C increase in upcoming years (Christensen et al., 2004), but precipitation projections are not as certain; many studies note the variability in precipitation trends, ranging from no change to reductions of about 10% basin-wide (Nash and Gleick, 1991, Christensen et al., 2004,Christensen and Lettenmaier, 2007 and Skoulikaris and Ganoulis, 2011). Warming in the basin has exceeded all other regions of the United States (NAS, 2007) and is projected to increase even more. Water resources of the UCRB are highly sensitive to snowfall and snowmelt timing and thus may be severely impacted by a changing climate.

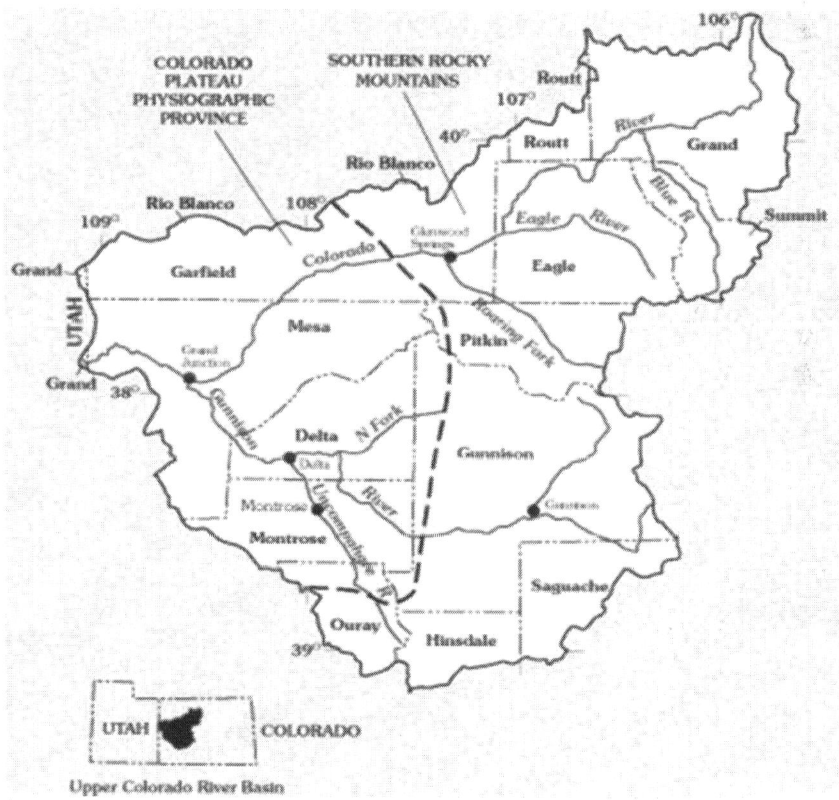

Figure 1. Upper Colorado River Basin, including Colorado and Gunnison Rivers.

Source: USGS NAWQA.

Over 75% of the streamflow is currently generated in the high elevation mountains from snowmelt (Nealon, 2008). A receding snowpack will alter runoff regimes and timing of peak spring flows, leading to more extreme events of flood and drought in both timing and frequency (Skoulikaris and Ganoulis, 2011). Regional water security and resiliency is determined by low-flows, which have already shown to be unsustainably low in summer especially with extreme flow conditions (both low and high) projected to increase under future climate scenarios.

Climate Models

The global climate models used in this study are three of the 24 identified by the Intergovernmental Panel on Climate Change (IPCC, 2007): Japan's Meteorological Research Institute's MRI-CGCM2.3.2 (MRI), Japan's MIROC3.2 (MIROC), and the Canadian Centre for Climate Modeling and Analysis CGCM3.1 (CGCM3) under Special Report on Emission Scenarios (SRES) A2 and B1 emission scenarios. The MRI is a coupled ocean-atmosphere GCM with a horizontal resolution of about 2.8 degrees (T42) with 30 vertical layers. The major parameterizations are cloud, convection, boundary layer, long and short wave radiation, wind, and temperature with recent improvements for radiation budget distribution along the meridians and improved the cloud radiative forcings. These parameterizations have improved reproduction of extreme events caused by climate change (Yukimoto, 2005). MRI also GHGs, CO_2, CH_4, N_2O, sulfate aerosols, and solar activity. The MIROC3.2 GCM is comprised of 5 components: atmosphere, land, river, sea ice, and ocean (K-1 Model Developers, 2004) with two atmospheric resolutions. We used the 2.8 degrees grid resolution with 20 levels (T42 L20). The CCCMa's CGCM3.1 is a coupled ocean-atmosphere GCM with a horizontal resolution of 2.8 degrees (T42) and 31 (L31) layers in the vertical. The major parameterizations are water vapor continuum (controlling infrared cooling rate), stomatal conductance, orographic gravity wave drag, cloud and solar properties (McFarlane et al., 2005) with updated representation of the water vapor continuum.

Climate Scenarios

In the SRES the IPCC identified six storylines for future emission pathways: (1) A1, which describes a future of rapid economic growth, a global population that peaks in mid-century and declines thereafter, and the introduction of new and more efficient technologies which include A1FI (a Fossil Intensive A1 scenario); (2) A1B (Emphasis on all energy sources); (3) A1T (Emphasis on non-fossil fuel energy sources); (4) A2, a Business-as-Usual (BAU) scenario that describes a very heterogeneous world with high population growth; (5) B1, which describes a world with the same low population growth as in A1 but with changes in economic structures toward a service and information economy; and (6) B2, which describes a world focused on local solutions to economic, social, and environmental sustainability. Of the six scenarios, A2 and B1 were selected for this study. They represent distinctly different paths of development. Each describes different demographic, social, economic, ecological, and environmental developments (Nakicenovic et al., 2000). Of the six, the A2 and B1 are most widely simulated in climate change studies (IPCC, 2007). They were chosen to represent climate change in this study because they describe a realistic range of conditions and uncertainties of the next century.

One of the primary factors that limit direct application of climate projections to hydrologic modeling for water resources assessment is the coarse spatial scale of GCM outputs. The grid resolution of GCMs used to develop future climate projections is at hundreds of kilometers, which is often difficult to relate to watershed-scale models. Thus, assessment of hydrologic impacts relies on spatial downscaling to translate the large-scale GCM projections to scales more representative of the physical implications of climate change (Christensen et al., 2004). There are two major techniques in downscaling: statistical and dynamic. Dynamic downscaling that simulates physical processes at finer scales is generally too computationally intensive for multi-decadal analysis (Maurer, 2007). Statistical downscaling is more widely used and essentially scales a GCM projection based on observed quantitative relationships between climates at the two spatial resolutions (Maurer et al., 2010). The three available methods of statistical downscaling considered for this study were as follows: bias-corrected and spatially downscaled (BCSD), constructed analogues (CA), and a hybrid method, bias-corrected constructed analogues (BCCA). Only BCCA was used in this study; however, general features of each method are described briefly below and illustrate why we implemented BCCA for this study.

There is usually a difference (bias) between observed and GCM-predicted climate outputs for almost all of the GCMs. In bias-corrected statistical downscaling, or BCSD approach, the observed climate data is evaluated against GCM output to remove bias. BCSD downscales monthly data, with a random resampling technique to generate daily values. This method inherently assumes that the climate projections simulations hold a constant bias in the GCM. The CA method, described by Hidalgo et al. (2008), uses daily large-scale output to directly downscale daily precipitation and temperature, instead of the more general monthly scale. It is based on the anomalies of daily precipitation and temperature but does not correct for bias. It ensures that daily fluctuations match the observed daily distribution, but does not guarantee monthly equivalents (Maurer et al., 2010). Furthermore, the same study found that in mountainous regions, the CA method produced better results, making it more appropriate to the UCRB.

The BCCA is a hybrid approach, following the CA approach of working with daily outputs but also recognizing the need for bias-correction. The process is almost identical to BCSD, but instead of applying it to monthly data, quantile mapping is imposed on daily values (Maurer et al., 2010). The BCCA technique explicitly corrects bias, creating datasets of actual values rather than anomalies (as in CA), but some downscaling bias remains. Capturing the variability of daily conditions is especially significant in a mountainous environment such as the CRB. Furthermore, the hydrology model used in this study (discussed below) requires daily data. Raff et al. (2009) demonstrated that while all downscaling methods produced reasonable streamflow statistics at most locations, the BCCA method consistently outperformed the other methods. BCCA captured daily large-scale skill and translated it to simulated streamflows with better reproduction of the observationally driven stream flows. Thus, the BCCA daily downscaled future data was selected to force a hydrologic model Watershed Analysis Risk Management Framework (WARMF). The BCCA daily datasets were recently made available by a joint project of the USBR, Lawrence Livermore National Laboratory (LLNL), Santa Clara University Climate Central, United States Geological Survey (USGS), and Scripps Institution of Oceanography. The hydrologic projection datasets were made possible by the efforts of the USBR, University of Washington's Climate Impacts Group (CIG) and the National Oceanic and Atmospheric Administration (NOAA) National Weather Service (NWS) Colorado Basin River Forecast Center. Their work created 1/8 degree gridded future climate hydrologic projection over the western U.S (Reclamation 2011, 2013). Time periods of historical (1961–2000), future Period 1 (2046–2065), and Future Period 2 (2081–2100) were used as these are the time scales for which the downscaled data is available for research use.

MATERIALS AND METHODS

Hydrologic Modeling and Model Development

Daily downscaled climate projections were used to drive the WARMF hydrologic model. WARMF was chosen for this study because it is a continuous simulation model that can provide a physical representation of the UCRB using a GIS database with inherent reservoir capabilities and stream diversions. A complete description is available in Chen et al. (2001). As a continuous model, WARMF considers soil moisture accounting and evapotranspiration. WARMF calculates daily runoff of a basin that is divided into catchments, stream segments, and lakes. Geza et al. (2009) successfully modeled the Turkey Creek Watershed of Colorado using WARMF. Their study analyzed calibration methods to improve performance of WARMF and was able to accurately simulate observed changes. WARMF uses daily time steps and requires daily records of precipitation, minimum and maximum temperature, cloud cover, dew point temperature, as well as air pressure and wind speed, which are used in the evapotranspiration, snow-formation, and snow-melt algorithms (Chen et al., 2001).

Daily precipitation is partitioned into rain and snow depending on daily temperature. Precipitation data is routed through the land layers to simulate snow and soil hydrology, producing runoff and shallow groundwater flow. A water-balance approach is used to calculate the hydrologic budget for each catchment resulting in runoff and base flow to river segments (Chen et al., 2001).

Several land use types can be defined for a sub-watershed-scale unit called a catchment (described in more detail below). The catchments may contain up to five soil layers. Each layer is assigned a thickness, initial soil moisture content, horizontal and vertical hydraulic conductivities, field capacity and porosity. The rate of percolation in the vadose zone from one soil layer to an underlying layer is limited by soil moisture content of the layer relative to its field capacity, level of saturation of the underlying layer, and vertical hydraulic conductivity. Lateral flow from a layer is based on Darcy's Law, where the head gradient is approximated as the slope of the land surface. The water balance is used to compute moisture content of each layer based on infiltration into the layer, percolation out of the layer, lateral inflow and outflow, and evapotranspiration (ET). Potential evapotranspiration is a function of total free surface water evaporation and soil transpiration based on Hargreaves equation. The model requires reservoir bathymetric data in the form of stage-area relationship.

The UCRB is delineated into 24 sub-watersheds, comprised of catchments, river segments, and 17 reservoirs (Fig. 2). The 24 sub-watersheds are further divided to over 600 catchments. All of the tributaries eventually feed into the main stem of the UCRB, prior to reaching the Colorado-Utah border. Model input parameters must be obtained for each catchment. Some input parameters are treated as catchment coefficients varied by catchment, defining the physical, chemical, and biological characteristics of each catchment, while others are defined as system coefficients with uniform coefficient applied to the entire basin.

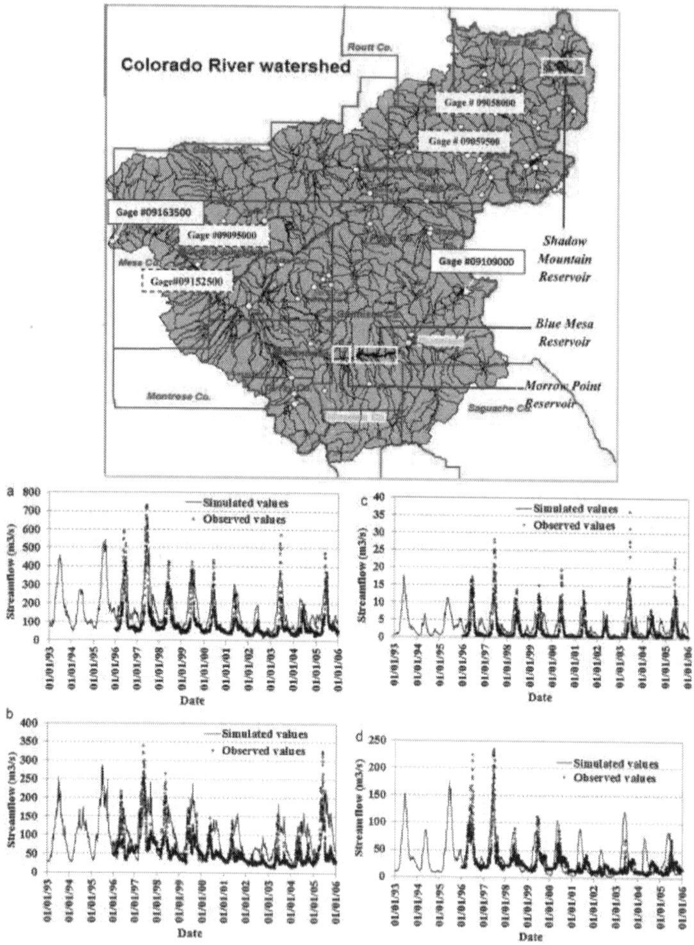

Figure 2. UCRB catchments, gauge stations, and reservoir locations. (a) Simulated versus observed flow at station #09095000 Colorado River near Cameo, CO (Low elevation station). (b) Simulated versus observed flow at station #09152500, Gunnison River near Grand Junction, CO (Low elevation station). (c) Simulated versus observed flow at station #09059500, Piney River near State Bridge, CO (High elevation station) station). (d) Simulated versus observed flow at station #0905800 at Colorado River near Kremmling, CO (High elevation).

Daily time-series of meteorological inputs such as precipitation and minimum and maximum temperatures are required for WARMF simulation. The data can be downloaded from the National Climactic Data Center (NCDC). USDA's National Resources Conservation Service (NRCS) Web Soil Survey (WSS) provides soil data and information. Soils data for the 13 counties were downloaded. The three prominent soil types in the basin were identified as sandstone (48%), limestone (15%) and 23 other soil types making the rest of the basin. Based on these soil types, soil parameters such as soil thickness, field capacity, porosity, horizontal and vertical conductivity were estimated. Land-use data were derived from the National Land Cover Dataset (NLCD) and imported to WARMF. The land surface of the basin is dominated by forest and rangeland. USGS water data web site (USGS, 2010) provides daily flow data for specific USGS gauging stations on river segments needed for model calibration. Diversions describe water removed from a stream predominantly for agricultural use. Diversions did not alter the water balance since most of the diverted water returns back to the streams as base or surface flow after use, over time. Our model included over 4500 diversions from Colorado Decision Support System (CDSS, 2010), which required an enormous effort to compile and post-process, but that greatly enhanced the realism of the model and improved model robustness. CDSS provides data organized by water division, river basin, and structure type. Our study area comprises of Colorado Water Divisions 4 and 5.

Seventeen reservoirs in the UCRB were incorporated into the WARMF model. The reservoirs that are used for hydropower include Blue Mesa, Morrow Point, and Green Mountain. Each reservoir is populated with observed data on reservoir geometry, existing water level, and stage-area-discharge data for the reservoir and the inflow/outflow structures. Physical data was obtained online from USBR (Reclamation, 2010) and through personal communication with each reservoir's operational personnel to develop stage-discharge tables describing spillway operation for reservoir simulation.

The impact on hydrology and reservoir storage due to climate change was analyzed under future climate projections using 3 GCMs and 2 SRES scenarios: MRI CGCM2 A2 and B1, CGCM3 A2 and B1, and MIROC A2 and B1. The output from each GCM was further categorized into two time frames: Period 1 (2046–2065) and Period 2 (2081–2100), with only Period 1 simulated in WARMF and compared with baseline scenario (1986–2000). Gauges were selected at significant locations considering watersheds, water supply impacts, climate characteristics, and hydropower influence (if leading to a producing reservoir), as well as geographic location (cross-boundary locations, elevation). Reservoirs were similarly selected, with most emphasis on hydropower production or potential at varying elevations.

WARMF Model Sensitivity Analysis and Calibration Using UCODE

Watershed models have more credibility when used as a decision-making tool or for assessing the impact of climate change after calibration. Geza et al. (2009), used UCODE (Poeter, 2005) for calibration of a WARMF model for Turkey Creek Watershed, Colorado. UCODE is a calibration and sensitivity analysis tool that minimizes the sum of weighted-squared-residuals with respect to the parameter values using a modified Gauss-Newton or Trust-Region method. We used UCODE linked to WARMF for sensitivity analysis and calibration. For the calibration, UCODE executes WARMF repeatedly, comparing simulated and observed data and adjusting parameter values to obtain the best fit.

Calibration of models with numerous input parameters requires identification of the most sensitive input parameters. Sensitivities reveal the significance of certain parameters on model performance. In UCODE, sensitivities are calculated and normalized to values between 0 and 1, with a value of 1 for parameters with maximum sensitivity and a value of zero for parameters that are not sensitive. Normally, parameters with a normalized value of less than about 0.01 are relatively insensitive and difficult to estimate during calibration and are associated with larger uncertainty (Hill and Tiedeman, 2007). It was determined through sensitivity analysis that the most sensitive parameters for the UCRB model are as follows: the Precipitation Weighting Factor (PF), a multiplier applied to the precipitation to account for local variations in precipitation amount from orographic effects; Evaporation Magnitude (EM), a scaling factor for evaporation used as an adjustment factor for ET calculated via the Hargreaves equation (Hargreaves, 1974); Evaporation Skewness (ES), a factor that accounts for seasonal variability of evaporation; Temperature Lapse rate (TF), a parameter that accounts for regional variations in temperature from orographic effects; Altitude Lapse rate (AF), a decrease in temperature with increasing elevation; and Soil Moisture Content (SM), Field Capacity (FC), and Horizontal Conductivity (Kh) (Table 1). Climate related parameters PF, EM, TF, AF and ES control the water balance and the remaining soil related parameters control runoff from rainfall.

Table 1. UCODE sensitivity analysis results of most significant parameters.

Parameter name	Parameter symbol	Normalized value	Literature value
Evaporation magnitude	EM	1	0.6–1.4
Evaporation skewness	ES	0.33	0.4–1.4
Temperature lapse rate (°C)	TF	0.31	−5 to 5
Precipitation weighting factor	PF	0.3	Varies by catchment (0.35–1.5)
Porosity or saturated moisture content	SM	0.2	0.2–0.6
Field capacity (m^3/m^3)	FC	0.1	0.0–0.4
Altitude lapse rate (°C/m)	AF	0.16	0.001–0.009

Other WARMF input parameters were found to be relatively not sensitive and were thus not included in detailed analysis.

Fig. 2 illustrates the gauge station locations used in calibration. In the study area, four gage stations were used to assess model fit (Figs. 2a–d). Three are located on the Colorado River. The first gage is Gauge #09095000, located at a low elevation on the Colorado River near Cameo. The other two gauges are further upstream at higher elevation on the Colorado River: Gauge #09058000, on the Colorado River near Kremmling, CO; and gauge #09059500, on the Piney River near State Bridge. The fourth gauge is located on the Gunnison River, Gauge #09152500 near Grand Junction.

The time series of observed vs. simulated values for gauges #09095000, #09152500, #09058000, #09059500 after model calibration are illustrated in Figs. 2a–d. We limited the calibration to only few sensitive parameters in order to achieve model convergence to a unique solution. Of the 9 parameters evaluated for system sensitivity, 3 were found to exhibit the most influence on model results. Calibration of a large hydraulic model to all 9 parameters was deemed too exhaustive and would not converge the model in a reasonable length of time. Only the most sensitive parameters of EM, ES, and PF were estimated during calibration. Literature values were used for the rest of the parameters. In WARMF, EM is a system coefficient and only one value could be used for all of the catchments in the basin.

The EM factor usually has a value between 0.6 and 1.4 (Herr et al., 2000) (see Table 1). With a starting value set at 1.0, the final calibrated value for EM was 1.3. In a region with much variable topography, such as the UCRB, the magnitude of evaporation is expected to vary greatly. Thus, ES was also calibrated to a value of 1.27 (within the literature range of 0.6–1.4). PF in WARMF is responsible for the effects of elevation difference on precipitation. PF was adjusted at the catchment level, based on site-specific conditions. Parameter values over all of the catchments varied from 0.35 to 1.5 and varying the PF factor by catchment during calibration significantly improved model performance.

Snow melt rates were generally less sensitive and were not critical to model performance. Open area melt rate is even less sensitive compared to forest area melt rates mainly due to limited area extent for open area in the basin. Hence melt rates were set to a value of 0.1 cm/day. Summary of parameter values determined via literature review and through calibration are listed in Table 1.

RESULTS

Precipitation Projections

We analyzed precipitation projections based on BCCA datasets at two contrasting locations in the UCRB with respect of elevation that are approximately 255 miles apart: a lower elevation station near the Utah State Line (4325 feet), and a higher elevation station near the Taylor Park Reservoir (9170 feet) (Fig. 3 and Fig. 4). Taylor Park reservoir sits along a tributary of the Gunnison River and serves as a good indicator of a possible high elevation impacts. The study demonstrated that projected hydrologic response to climate change is greatly influenced by elevation. The figures illustrate the mean daily precipitation (inches), organized by month, based on average daily accumulation for both A2 and B1 (using all 3 GCMs) climate scenarios for a specific time period.

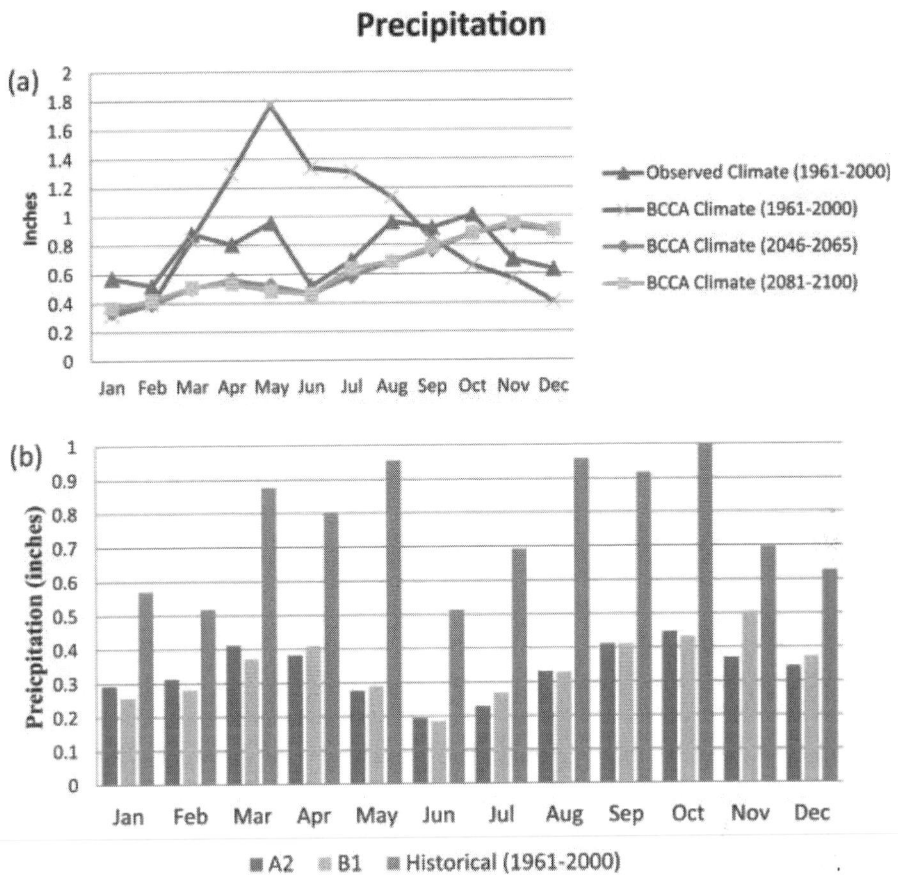

Figure 3. (a) Average of daily precipitation, summarized for each month, at low elevation near Utah State Line (USGS #09163500). (b) Average of daily precipitation summarized for each month at low elevation near Utah State Line under SRES A2 & B1 of all GCMs in Period 1 (2046–2065) compared to observed climate data.

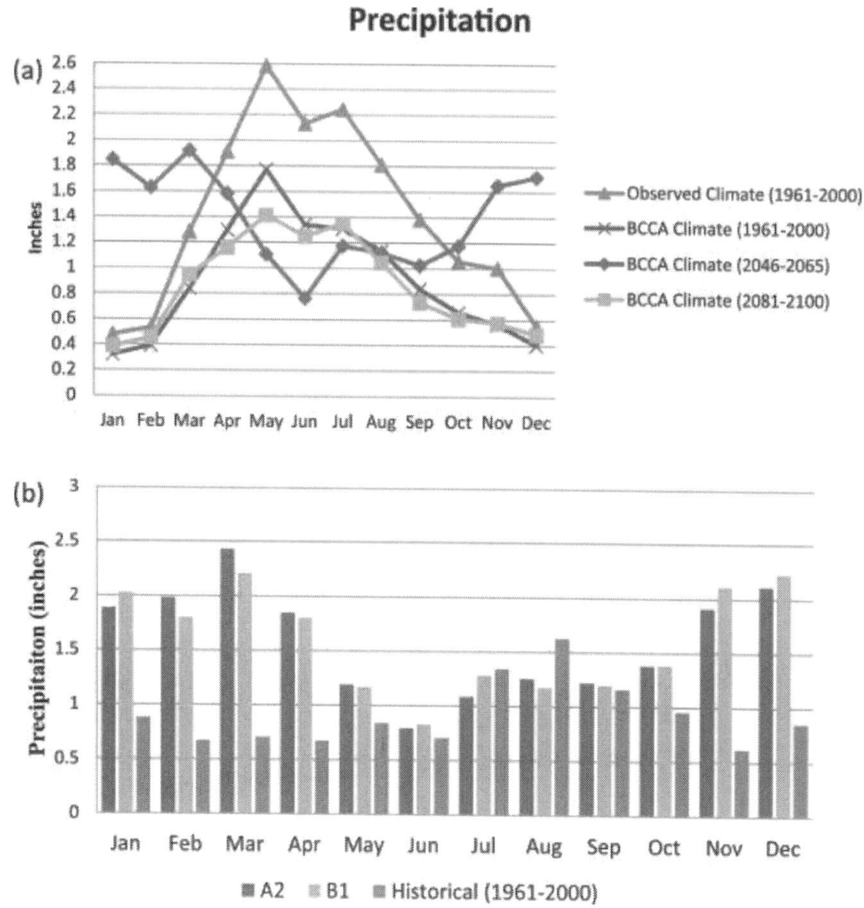

Figure 4. (a) Average of daily precipitation, summarized for each month, at high elevation near Taylor Park Reservoir (USGS #09109000). (b) Average of daily precipitation at high elevation near Taylor Park Reservoir summarized for each month under SRES A2 & B1 of all GCMs in Period 1 (2046–2065) compared to observed climate data.

Fig. 3 and Fig. 4 illustrate precipitation trends exclusive of all climate scenarios and models, isolating only the time period. The variability across the climate models and the two scenarios was relatively small compared to the difference between historical data and GCM predictions; thus, we compared the observed values to the average of all climate models and scenarios. The purpose being to demonstrate performance of GCMs compared to observation, the figures include the historical precipitation values (1961–2000) and precipitation predictions from downscaled GCM data using BCCA method for the same historical period. At low elevation, historical precipitation prediction from BCCA captured seasonal variation in the observed data but differed in magnitude. At higher elevation, precipitation predictions from BCCA illustrate a trend toward increased precipitation for the winter months, but a decrease in spring months. The analysis showed the model captured the general seasonal trend but the observed data and the climate model showed a considerable difference in magnitude with BCCA over-predicting precipitation during the high rainfall months in the lower elevation and consistently under-predicting precipitation in the higher elevation. Our results showed that for lower elevation, the climate projections describe a reduction in annual precipitation (compared to historical observed values) illustrated in Fig. 3 and Fig. 4. Assuming the over-predictions at low elevation are propagated to future years, the lower elevations would have even higher shortage of precipitation and reduction in stream flows than predicted in our study. Oppositely, assuming the BCCA under-predicted future climate as was seen in the observed data, the higher elevations would experience increases in precipitation and increased runoff. This supports prior work which emphasized wetter conditions during winter months and drier in summer in the near future (2046–2065), but overall drying toward the end of the 21st Century. The GCMs predicted reductions in precipitation (10–40%) compared to observed values at lower elevations, while at higher elevation an increase in precipitation (20–35%) was observed during some months.

Fig. 3 and Fig. 4 describe impacts of the SRES scenarios A2 and B1 on predicted precipitation under each GCM at low elevation. For lower elevation, the climate projections describe a reduction in annual precipitation (compared to historical observed values) of over 54%. Similarly, at low elevation, the largest reduction in future precipitation (greater than 60%) occurs during the summer months of May to August. This reduced precipitation combined with increasing summer temperatures (discussed below) is expected to exacerbate the drying conditions and reduction in stream flow. From November to February, the reduction in precipitation is relatively lower, although specific results vary depending on climate scenario. Low elevations under B1 scenario show a reduction in November precipitation of (only 27% less than the historical), but the reduction becomes 47% under the A2 condition. January and February amounts also differ depending on climate scenario, with A2 showing more precipitation during both months. These finding echo prior climate studies that indicate wetter winters and hotter, drier summers, under extreme climate conditions.

On the contrary, the higher elevation (Fig. 4a) experiences precipitation volume increase from 57% to 71% (depending on climate scenario) compared to historical data, but distinct drying during summer (May to August), when future precipitation falls below historic values. The greatest precipitation increases under future climate scenarios occur during the months of December and March. In mountainous high elevation terrains such as the one in our study, the months of December through March experience the most frequent and greatest snow accumulation largely due to significant orographic effect. During the summer drying period, similar to that of lower elevation, a reduction in precipitation is predicted compared to historical data (as much as 27%) (Fig. 4b). Again, these drying conditions are exacerbated by predicted temperature changes in the region (discussed in Section 3.2). This agrees with the IPCC's (2007) finding of substantial summer drying in the mid-latitudes. These trends, although averaged across the three GCMs, display a strong dependency on elevation.

We observed that climate change is closely linked to elevation in the UCRB. Thus, changes in reservoir operations and management in response to climate change should be addressed based on reservoir elevation and location. Many of the reservoirs on the UCRB, including those with the most hydropower production potential, reside at high elevations. From these results, we can see that there will be significantly more water in winter months from precipitation increase, but severe drying in summer. This is problematic because water shortages generally occur during summer months. This will require adaptive policies and action by water managers (presented in more detail in a subsequent section).

Next, we analyzed the differences between GCMs with respect to precipitation predictions in only the low elevation (Fig. 5 and Fig. 6) station to gain a better understanding of how variability in climate models might impact the results of a hydrologic model intended to analyze climate-change impacts on water resources. The B1 scenario shows 8% more precipitation annually than the A2 scenario. Winter precipitation is more pronounced under A2, as is summer drying. The A2 scenario also exhibits more extreme seasonality than B1. Under both the A2 and B1 scenarios, CGCM3 trends toward higher precipitation values, MIROC generally describes the lower values, and the MRI predictions are intermediate.

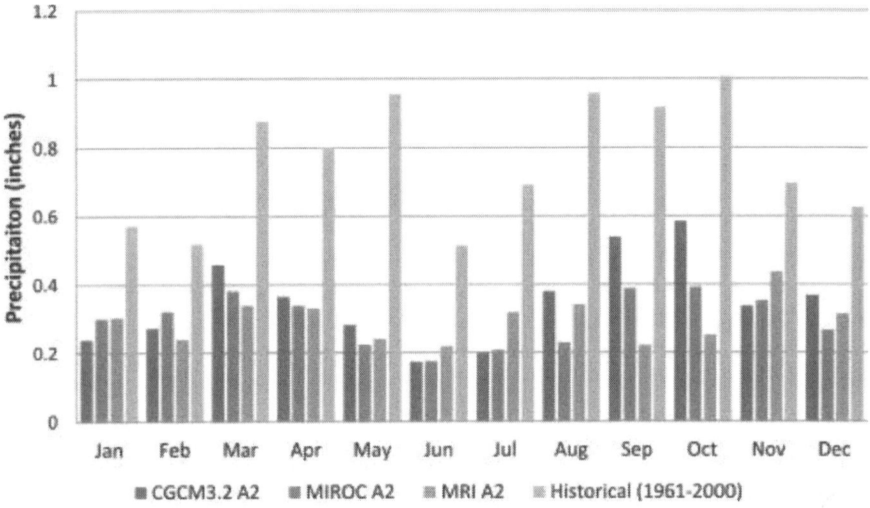

Figure 5. Daily-averaged precipitation comparison of GCMs under SRES A2 condition in Period 1 (2046–2065) and observed historical climate at low elevation near Utah State Line.

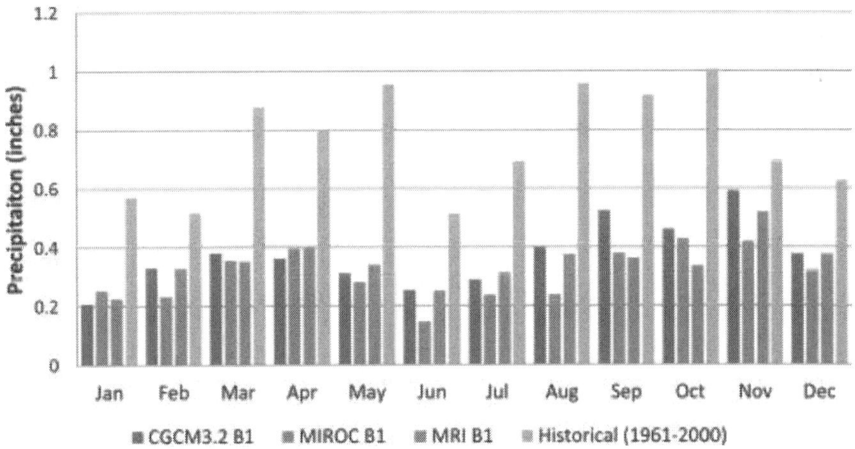

Figure 6. Daily-averaged precipitation comparison of GCMs under SRES B1 condition in Period 1 (2046–2065) and observed historical climate at low elevation near Utah State Line.

These results suggest a consistent bias exists between the different GCMs, but the difference was not more than 16%. When analyzing GCM results for climate change analysis, no single model can be regarded as an ideal one. Thus, the major purpose and benefit of analyzing several climate models, as in this study, is to produce a range of outputs from different GCMs and investigate the envelope of influence in climate projections on the water resources, such as predicted stream flow and reservoir levels, to better understand associated uncertainties. This approach should aid water managers and operators in assessing future changes and risks under uncertainty.

Temperature Projections

Fig. 9 and Fig. 10 describe the average of daily minimum temperatures summarized for each month at the Utah State Line (low elevation) and Taylor Park (high elevation) gauges. At low elevation (Fig. 9), GCM predictions describe an increase in minimum temperatures compared to historical observations for the historical period (1961–2000). The increase is more pronounced for the far future, Period 2 (2081–2100), compared to near future, Period 1 (2046–2065). This behavior differs from precipitation trends, where the decline was more pronounced in the near future of Period 1 (2046–2065) than the far future (2081–2100). It is widely agreed that global temperatures will increase under future climate conditions (Christensen and Lettenmaier, 2007 and Barnett and Pierce, 2009; and others). There is a general trend toward warming at high and low elevations, but the increase in temperature is predicted to be greater at lower elevations, exacerbating already dry and warm conditions. Fig. 9 illustrates the months of January through March experiencing higher temperatures under observed climate, rather than predicted. With a reduction in minimum temperatures in the future, snow as snow and not as melt may propagate for longer than has been expected in the historical past at higher elevations. At low elevations, the opposite is true, where the minimum temperature has been shown to increase in winter and generate early snow melt, specifically an increase in early spring flows (1–2 weeks earlier) than in the past (IPCC, 2007). Fig. 7 illustrates a 3–5 degree Fahrenheit (°F) mean annual increase in maximum temperature at low elevation, while Fig. 8 shows an increase of 0.7–2 °F at higher elevation.

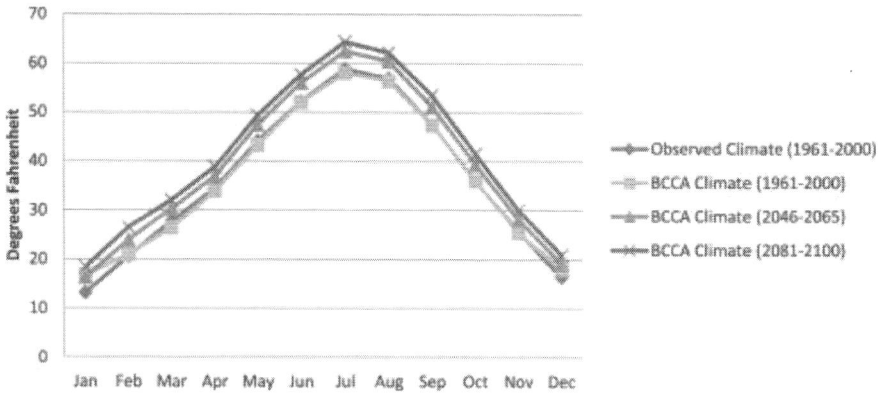

Figure 7. Average of Daily Maximum Temperatures (F) for each month under Future Climate at #09163500 at low elevation near Utah State Line for future time periods.

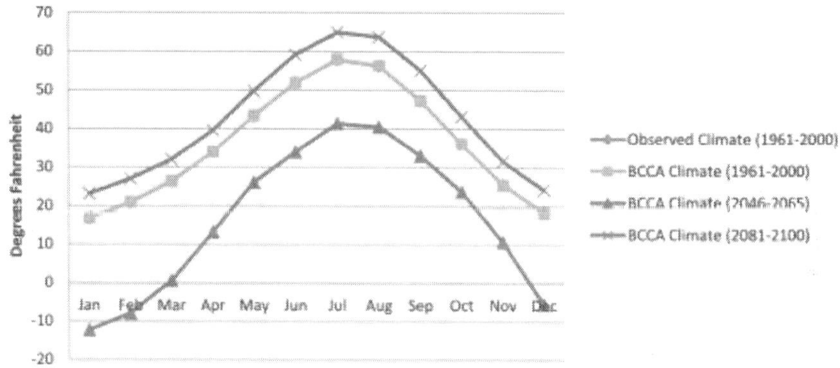

Figure 8. Average of Daily Maximum Temperature (F) for each month under Future Climate at #09109000 at high elevation near Taylor Park for future time periods.

Temperature changes in response to climate scenarios A2 and B1 are illustrated in Fig. 9 and Fig. 10 for lower and higher elevation, respectively, during Period 1 (2046–2065). The maximum change in temperature at lower elevation is 3.5 °F for A2 scenario and 2.8 °F for B1 scenario, as compared to historical data. Rise in temperature is exacerbated during the summer months of June to September, where a 4 °F difference is produced. The difference between the two climate scenarios, A2 and B1, was only 1.8% (or 0.7 °F), showing the uniformity of increasing summer temperatures. At high elevations, temperature is also predicted to increase but differs in timing and is much less pronounced in magnitude. Fig. 10 illustrates the pattern of winter months growing colder, while the summer months become hotter. A2 illustrates only a maximum increase of 2 °F and B1 demonstrates a 1.5 °F rise, as compared to historical data. With temperatures actually decreasing in winter months (as much as 9 °F decrease in March), this directly impacts snowmelt and, consequently, streamflow. Hydropower is inherently dependent on streamflow, which is highly influenced by snowmelt at high elevations. The timing of snowmelt would be altered at high elevations under future temperature conditions, due to changes in the hydrograph, with melt occurring earlier in the season than historical averages.

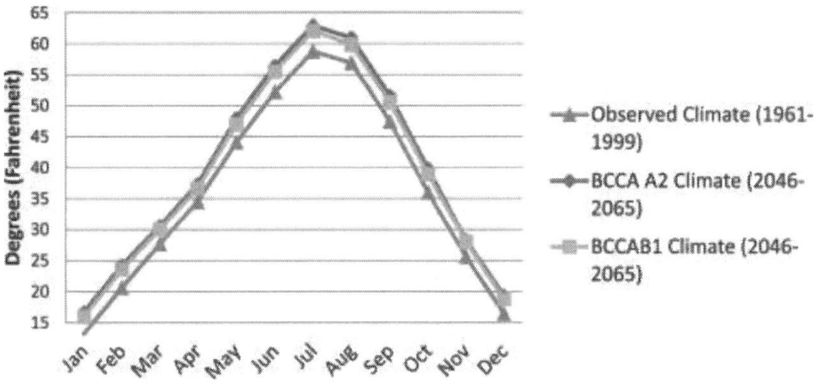

Figure 9. Comparison of minimum temperature at low elevations for scenarios A2 and B1 with observed climate Period 1 (2046–2065) near Utah State.

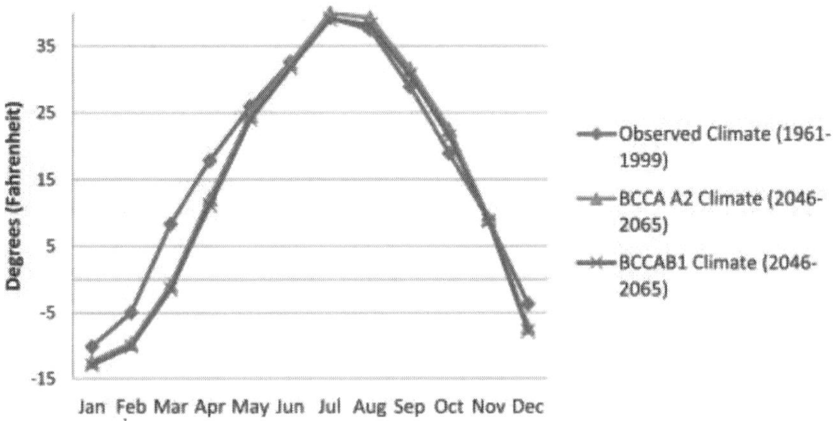

Figure 10. Comparison of minimum temperature at high elevations for scenarios A2 and B1 with observed climate Period 1 (2046–2065) near Taylor Park.

Stream Flow Projections

The WARMF hydrologic model was used to analyze the impact of the projected climate changes presented earlier on water resources in the UCRB. Fig. 11, Fig. 12 and Fig. 13illustrate stream flow changes under scenarios A2 and B1 for the 3 GCMs at flow gauge near the Utah state line, after the confluence of the Gunnison and Colorado Rivers. Historical stream flow values are significantly greater than any of the predicted scenarios, showing a pronounced reduction in future stream flow. Future stream flow is shown to decline considerably by about 62%. Most pronounced is the change in summer months (June to August), followed by September and October months. Historical peak flows generally occurred between May and July, although distinct peaks were observed during May and June as shown in Fig. 11. This moves the centroid of the hydrograph to an earlier period of April to June from the observed May to July period under historical conditions, resulting in most of the stream flow to occur a month earlier. In Fig. 12, stream flow projections are compared for the different GCMs, at low elevation near the Utah-Colorado State line. Projected stream flows are 52% greater under CGCM3 than MIROC (similar to precipitation trends) in the B1 scenario, but only 4% more in A2 conditions. MRI results illustrate 50% more streamflow under A2 conditions than in B1. MIROC behaves similarly to MRI in A2 scenario but is only half of the B1 predicted streamflow for both MRI and CGCM3. Essentially, MIROC consistently shows the least streamflow (consistent with our prior findings), but not the least precipitation accumulation; hence, streamflow response is not linearly dependent on precipitation trends. Although MIROC predicts an overall decrease in streamflow, it does show higher volumes in the summer than MRI. Overall, all GCMs illustrate an average streamflow reduction of approximately 78% in A2 and 72% in B1, with the greatest decrease described in the months of May through August. From these results, we can see that hydrologic response may differ based on the GCM used to predict the changes but the impact is the same: an overwhelming reduction in future streamflows along the Colorado River.

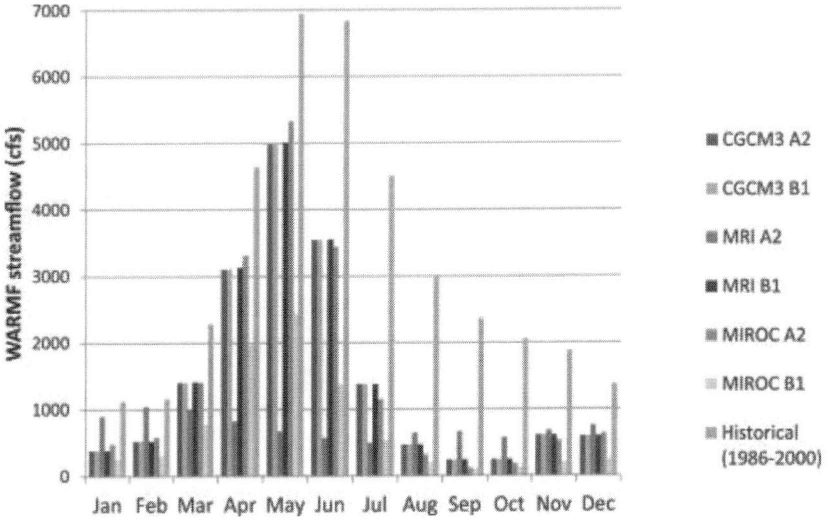

Figure 11. Comparison of predicted streamflow under all GCMs and emission scenarios in Period 1 (2046–2065) at gage #09163500 at low elevation near Utah State Line.

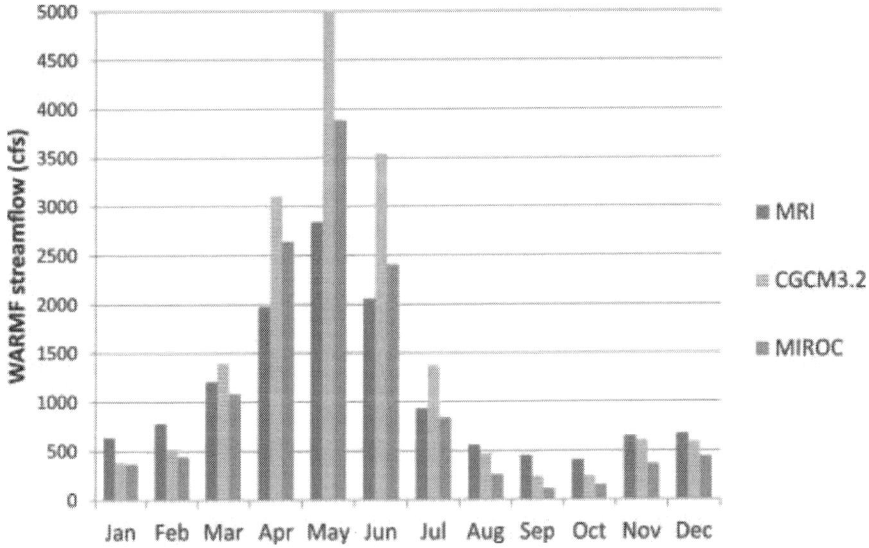

Figure 12. Comparison of predicted streamflow (cfs) for each GCM, average of A2 and B1 scenarios in Period 1 (2046–2065) at low elevation near Utah State Line.

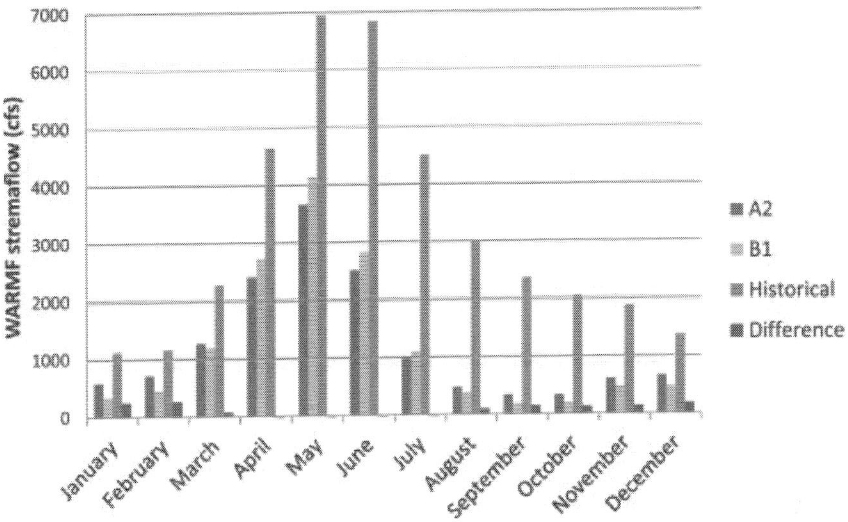

Figure 13. Mean of daily streamflow values from SRES A2 and B1 compared historical values. The difference value is between A2 and B1 in Period 1 (2046–2065) at low elevation near Utah State Line.

Fig. 13 describes the impact of the emission scenarios A2 and B1 on stream flow in comparison to historical values and to each other (difference). A2 illustrates a more extreme behavior envelope, i.e., more wet winters and more dry summers. For the months of November to February, streamflow under scenario A2 is greater than that simulated by the B1 scenario. However, streamflow is dramatically reduced in rainfall–runoff summer months (April to July) under A2. Temperature is also predicted to increase annually, resulting in increased evaporation rates and reduced soil moisture, and therefore reduced runoff; flow is shown to decrease during the most affected summer months under both A2 and B1 scenarios. A 62% reduction occurs in annual flow volume occurs under A2, exacerbating the already over-allocated water deliveries downstream.

Reservoirs and Hydropower Production Impact

The hydrologic impact of climate shift on reservoir levels and hydropower production was analyzed considering an overall increase in temperature and flow reduction, with earlier snowmelt across the UCRB. Hydropower-producing reservoirs Blue Mesa, Morrow Point, and Shadow Mountain Reservoir (Fig. 2) were assessed. The forthcoming discussion focuses on comparison and evaluation of monthly averaged historical reservoir volume and spill versus the predicted volumes for scenarios A2 and B1 (averaged across the three climate models). Spill is the volume released without producing hydropower when river flows exceed the hydraulic capacity of the system to generate electricity. The Blue Mesa reservoir is the largest reservoir in the state, at an elevation of more than 7500 feet with a surface area of 9200 acres. It is located on the Gunnison River and produces over 260 MW each year. Under projected climate conditions, it experiences reduction of 76–85% in spill (depending on scenario and GCM) and 70–73% in volume (Fig. 14). Note that peak storage occurs two months earlier (in May instead of the historical July) for both emission scenarios, which is attributed to earlier snow melt causing earlier peak in volumes. Despite a projected temperature increase, a decrease in evaporation is expected (not shown), which is attributed to the large decrease in surface area. The decrease in surface area is attributed to depletion in storage volume due to increasing temperatures in summer months which is propagated to the lower water levels, thus decreasing surface area from which evaporative losses become less influential.

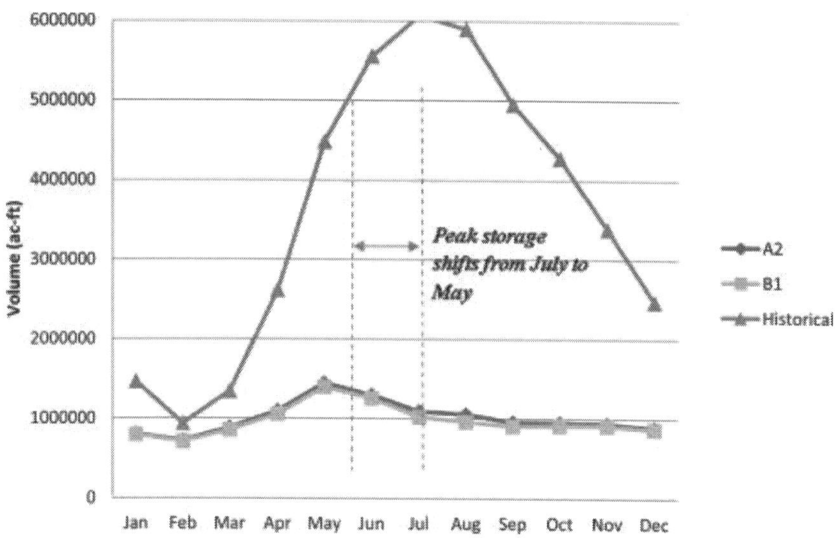

Figure 14. Monthly Mean Water Storage for all GCMs for Blue Mesa Reservoir in Period 1 (2046–2065).

Morrow Point reservoir is 12 miles downstream of Blue Mesa on the Gunnison River with a surface area of 918 acres and a maximum capacity of 117,000 acre-feet. Its 173 MW capacity is used for water supply and irrigation. Under future climate projections, changes in spill and volume were not nearly as significant as for Blue Mesa (Fig. 15), although a general trend indicated slightly reduced reservoir levels in the winter months, with increased levels during the late spring and summer months. The projected climate, elevation, and location of this reservoir are not much different than that of the Blue Mesa Reservoir, but the two vary greatly in size. Blue Mesa reservoir has a maximum capacity of 940,700 acre-feet and is located about 350 feet higher than Morrow Point. The relatively small change in Morrow Point storage is attributed to its being managed by Blue Mesa Reservoir upstream; the larger reservoir controls the flows to Morrow Point, mitigating climate influenced impacts of increasing temperatures and decreasing precipitation. Since Blue Mesa controls the flow, its reduction in volume due to a reduction in precipitation and increase in temperature also results in less flow to Morrow Point and decrease in storage volume. This demonstrated that there is a cross basin effect between reservoirs under climate change scenarios which could be different from what is observed based on historical climate, indicating the need for updating of reservoir rule curves considering the impact of climate change.

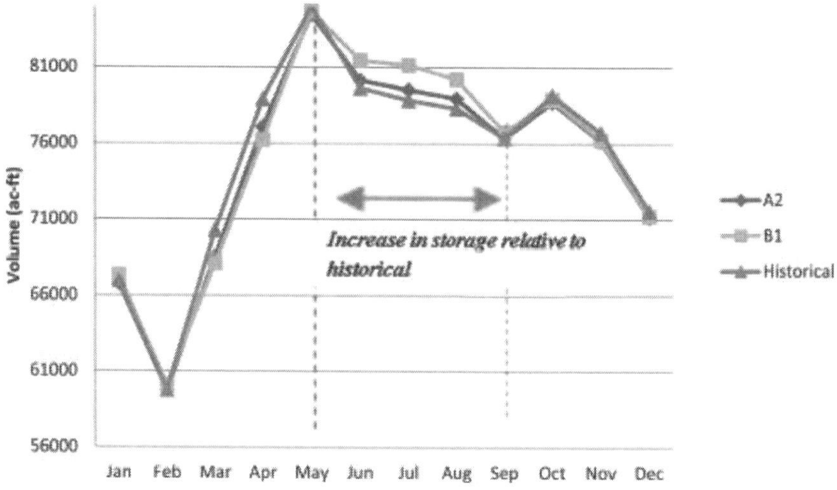

Figure 15. Monthly Mean Water Storage for all GCMs for Morrow Point Reservoir in Period 1 (2046–2065).

Shadow Mountain reservoir lies upstream of the glacial limit at about 8300 feet, located in a glacial moraine. It is part of the Colorado-Big Thompson project water collection system and is critical to providing water for consumption, energy, and irrigation needs. It is approximately 1337 surface acres with maximum capacity of 17,354 acre-feet. Under future climate conditions, there is a 1–3% increase in precipitation in the proximity of the reservoir and a 5–7% reduction in overall evaporation. The hydrologic model predicts 8–10% increase in spill and a 1–5%

increase in total volume (Fig. 16). A slight decrease in volume is predicted during late winter and spring, with a corresponding slight increase in volume from June to September under both climate scenarios, whereas a slight increase is projected from September to December for the reduced emission (B1) scenario. The earlier onset of snowmelt and subsequent earlier intense runoff may also have a significant impact on hydropower, particularly for small reservoirs. Earlier more intense runoff leads to more dam spillage during peak storage, reducing the volume retained entering the summer months when power demand is highest. Additional storage for early spring melt in smaller reservoirs is necessary to avoid spillage in spring to enable operators to accommodate demand later in summer.

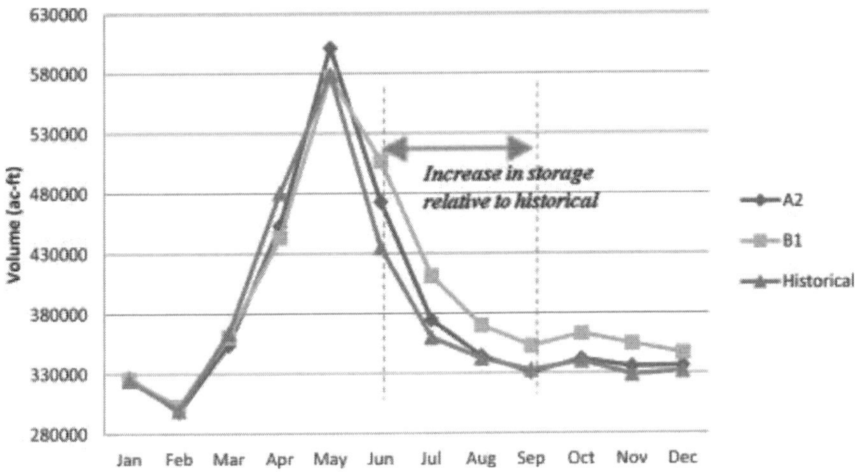

Figure 16. Monthly Mean Water Storage for all GCMs for Shadow Mountain Reservoir in Period 1 (2046–2065).

Hydropower production is inherently reliant on storage or head for energy production. In the above scenarios, power generation for the Blue Mesa Reservoir could decline significantly due to reduced volume and head. Adaptive management is necessary to address a changing environment. This involves identification of varying climate variables (precipitation, temperature) and implementing strategies to mitigate the variability. Then, perhaps most significantly, we can evaluate the effectiveness of the new strategies and adapt them further if results are unsatisfactory. A significant portion of adaptive management involves changing existing rule curves, which are used to guide reservoir operations with regard to power generation. Rule curves describe the water level, or range of acceptable levels, for each day of the year (i.e., levels needed to generate a certain hydraulic head and thus control power generation to meet historic demand on a daily basis). Reservoir rule curves specify the target elevation of the reservoir. When water level is below the level specified by the rule curve, discharge to downstream could be reduced to increase water level in the reservoir and release as much as possible to regain flood storage when storage levels exceed the rule curve. The amount that you release can be constrained by factors such as downstream flooding and the physical release capacity at the dam. Dropping the rule curve to a lower elevation will produce more flood storage; however, those decisions and established practices were set based on historic levels and approaches. Rule curves need to be accurately managed as they may optimize flood operations but negatively impact other operations such as water supply and hydropower production. Furthermore, rule curves need to be adapted to the changing climate, as levels that were historically acceptable and annual time periods for which they were created may no longer apply in future climate conditions.

CONCLUSIONS

Future projections show that precipitation changes vary by elevation. At low elevation near the Utah state line, the projections describe more than a 60% decline in precipitation. Conversely, at high elevation near Taylor Park Reservoir projected precipitation increases by as much as 74% compared to historical values. Regardless, all elevations depict a decline during summer months, which agrees with the IPCC's (2007)finding of substantial summer drying in the mid-latitudes. A basin-wide increase in temperature, especially the minimum temperature, causes precipitation to fall as rain instead of snow. At lower elevations, the projections show an increase in temperature more pronounced during the summer months of June to September. At high elevations, temperature is also predicted to increase but differs in timing and is much less pronounced. Under the higher emission A2 scenario, streamflow is greater during winter months but is dramatically reduced in runoff-dominated spring and summer months (April to July). Under A2 scenario, the hydrologic response is more pronounced when precipitation effects are strongest in winter. Temperature is projected to increase annually resulting in increased evaporation and reduced soil moisture (the latter results in greater infiltration capacity during storms), leading to decreased stream flow during the summer months under both A2 and B1 scenarios. This process exacerbates and/or creates drying conditions. With an increase in temperature of 2–4 °F basin-wide, timing and magnitude of stream flow is also impacted. We found snowmelt to occur 2–4 weeks earlier, leading to reduction in streamflow.

Our simulation results showed reduction in water storage at the high elevation for hydropower-generating Blue Mesa reservoir, while only slight changes are projected 12 miles downstream at Morrow Point reservoir. This discrepancy in water storage is attributed to the influence of surface area on evaporative demand, where the latter reservoir was less impacted by evaporation. At Shadow Mountain reservoir, volume increased by up to 5%, mainly ascribed to precipitation increases from orographic effects.

Hydropower is dependent on inflow conditions. Smaller reservoirs are most significantly impacted if early melt inflow overwhelms capacity; necessitating spill of extra water needed in later summer months. Additional spilling of reservoirs in early spring could limit water volume in reservoirs during summer and make it more difficult for power operators to meet peak demand. Larger reservoirs are less impacted but still experience a changed regime in response to climactic changes. Water managers will need to adjust rule curves and operation policies to accommodate all demands (specifically for multi-use reservoirs of water supply, irrigation, and hydropower production).

Projected warmer and drier conditions resulted in reduction of river flow and increased evaporation, which tend to reduce reservoir levels and thus hydropower capacity but also impact the ability to meet water subscriptions downstream. Most of the sub-catchments of the UCRB face a decrease in future precipitation, with potential for more frequent occurrence of intense events. The combination of less precipitation with higher temperatures indicates a potential for water stress. Adaptive policies are necessary to deal with projected changes. Without proactive action, current practices will lead to annual loss in hydropower potential and reliability in future climate. Rule curves, on which operations are based, must be changed to optimize production while saving water when necessary and minimizing flood risk.

REFERENCES

1. Barnett, T.P., Pierce, D.W., 2009. Sustainable water deliveries form the Colorado River in a changing climate. Proc. Natl. Acad. Sci. U. S. A. 1106 (18), 7334–7338.

2. CDSS, 2010. Colorado Decision Support Systems, Retrieved from http://cdss.state.co.us/Pages/CDSSHome.aspx
3. Chen, C.W., Herr, J.W., Weintraub, L., 2001. Watershed Analysis Risk Management Framework (WARMF): Update One – A decision support system for watershed analysis and total maximum daily load calculation, allocation and implementation. Publication No. 1005181. Electric Power Research Institute, Palo Alto, CA.
4. Christensen, N.S., Wood, A.W., Voisin, N., Lettenmaier, D.P., Palmer, R.N., 2004. Effects of climate change on the hydrology and water resources of the Colorado River Basin. Clim. Change 62, 337–363.
5. Christensen, N.S., Lettenmaier, D.P., 2007. A multimodel ensemble approach to assessment of climate change impacts on the hydrology and water resources of the Colorado River basin. Hydrol. Earth Syst. Sci. 3, 3727–3770.
6. Geza, M., Poeter, E.P., McCray, J.E., 2009. Quantifying predictive uncertainty for a mountain-watershed model. J. Hydrol. 376 (1–2), 170–181, http://dx.doi.org/10.1016/j.jhydrol.2009.07.025
7. Hargreaves, G.H., 1974. Estimation of potential and crop evapotranspiration. Trans. ASAE 17, 701–704. Herr, J., Weintraub, L., Chen, C., 2000. User's Guide to WARMF: Documentation of Graphical User Interface. EPRI, Palo Alto, CA (Report EP-P2346/C, p. 1054).
8. Hidalgo, H.G., Dettinger, M.D., Cayan, D.R., 2008. Downscaling with constructed analogues: daily precipitation and temperature fields over the United States. California Energy Commission, Public Interest Energy Research Program, Sacramento, CA, p. 62.
9. Hill, M.C., Tiedeman, C.R., 2007. Effective Groundwater Model Calibration, with Analysis of Sensitivities, Predictions, and Uncertainty. Wiley and Sons, New York, pp. 455. IPCC, 2007. Climate Change 2007: The Scientific Basis. In: Solomon, S., Qin, D., Manning, M., Chen, Z., Marquis, M., Averyt, K.B., Tignor, M., Miller, H.L. (Eds.), Contribution of Working Group I to the Fourth Assessment Report of the Intergovernmental Panel on Climate Change. Cambridge University Press, Cambridge, United Kingdom/New York City, United States. K-1 Model Developers, H.H., 2004.
10. K-1 coupled model (MIROC) description, K-1 technical report. University of Tokyo, Tokyo. Maurer, E.P., 2007. Uncertainty in hydrologic impacts of climate change in the Sierra Nevada, California under two emissions scenarios. Clim. Change 82 (3–4), 309–325, http://dx.doi.org/10.1007/s10584-006-9180-9
11. Maurer, E.P., Hidalgo, H.G., Das, T., Dettinger, M.D., Cayan, D.R., 2010. Assessing climate change impacts on daily streamflow in California: the utility of daily large-scale climate data. Hydrol. Earth Syst. Sci. 7, 1209–1243. McCabe, G.J., Wolock, D.M., 2007. Warming may create substantial

water supply shortages in the Colorado River basin. Geophys. Res. Lett. 34, L22708, http://dx.doi.org/10.1029/2007GL031764

12. Miller, W.P., Piechota, T.C., Gangopadhyay, S., Pruitt, T., 2011. Development of streamflow projections under changing climate conditions over the Colorado River basin headwaters. Hydrol. Earth Syst. Sci., 2145–2164, http://dx.doi.org/10.5194/hessd- 7-5577-2010
13. Nakicenovic, N., Alcamo, J., Davis, G., de Vries, B., Fenhann, J., Gaffin, S., Gregory, K., Grubler, A., Jung, T.Y., Kram, T., La Rovere, E.L., Michaelis, L., Mori, S., Morita, T., Pepper, W., Pitcher, H., Price, L., Riahi, K., Roehrl, A., Rogner, H.-H., Sankovski, A., Schlesinger, M., Shukla, P., Smith, S., Swart, R., van Rooijen, S., Victor, N., Dadi, Z., 2000. IPCC Special Report on Emissions Scenarios. Cambridge University Press, Cambridge, United Kingdom/New York, NY.
14. McFarlane, N.A., Scinocca, J.F., Lazare, M., Harvey, R., Verseghy, D., Li, J., 2005. The CCCma third generation atmospheric general circulation model. CCCma Internal Report, pp. 25.
15. Nash, L.L., Gleick, P.H., 1991. Sensitivity of streamflow in the Colorado Basin to climatic changes. J. Hydrol. 125, 221–241.
16. Nealon, T., 2008. Sensitivity analysis and calibration of a surface water model of the Upper Colorado River Watershed. Colorado School of Mines, pp. 106.
17. Poeter, E.P., 2005. UCODE 2005 and Six Other Computer Codes for Universal Sensitivity Analysis, Calibration, and Uncertainty Evaluation, Golden, CO, United States. Raff, D.A., Pruitt, T., Brekke, L.D., 2009. A framework for assessing flood frequency based on climate projection information. Hydrol. Earth Syst. Sci. 13, 2119–2136, http://dx.doi.org/10.5194/hess-13-2119-2009
18. Rasmussen, R., Liu, C., Ikeda, K., Gochis, D., Yates, D., Chen, F., Tewari, M., Barlage, M., Dudhia, J., Yu, W., Miller, K., Arsenault, K., Grubisiˇ c, ´ V., Thompson, G., Gutmann, E., 2011. High-resolution coupled climate runoff simulations of seasonal snowfall over Colorado: a process study of current and warmer climate. J. Clim. 24, 3015–3048, http://dx.doi.org/10.1175/2010JCLI3985. 1
19. Skoulikaris, C.G., Ganoulis, J., 2011. Assessing climate change impacts at river basin scale by integrating global circulation models with regional hydrological simulations. Eur. Water 34, 55–62.
20. USGS, 2010. USGS Surface-Water Data for the Nation, Retrieved from http:\\waterdata.usgs.gov/nwis/sw U.S. Reclamation, 2010. Reservoir Operations, Retrieved from http://www.usbr.gov/uc/wcao/water/index.html
21. Yukimoto, S.U., 2005. Model Information of Potential Use to the IPCC Lead Authors and the AR4. Japan.

CITATION

M. Kopytkovskiy, M. Geza, J.E. McCray, Climate-change impacts on water resources and hydropower potential in the Upper Colorado River Basin, Journal of Hydrology: Regional Studies, Volume 3, March 2015, Pages 473-493, ISSN 2214-5818, http:// dx.doi.org/10.1016/j.ejrh.2015.02.014.

CHAPTER 7

Climate Risk Assessment for Water Resources Development in the Niger River Basin Part I: Context and Climate Projections

J. G. Grijsen[1], C. Brown[2], A. Tarhule[3], Y. B. Ghile[4], Ü. Taner[5], A. Talbi-Jordan[6], H. N. Doffou[7], A. Guero[7], R. Y. Dessouassi[7], S. Kone[7], B. Coulibaly[7] and N. Harshadeep[8]

[1]Independent Hydrology and IWRM Consultant, Texas, USA
[2]Department of Civil and Environmental Engineering, University of Massachusetts, Amherst, USA
[3]Department of Geography and Environmental Sustainability, University of Oklahoma, Norman, USA
[4]Woods Institute for the Environment, Stanford University, Stanford, USA
[5]Dept. of Civil and Environmental Engineering, University of Massachusetts, Amherst, USA
[6]The World Bank Middle East North Africa (MNSWA), Washington, USA
[7]Niger Basin Authority/ Autorité du Bassin du Niger (ABN), Niamey, Niger
[8]The World Bank, Africa Region, Washington DC, USA

INTRODUCTION

Climate change has emerged as one of the most important challenges of the 21st century. Consequently, a plethora of studies has appeared which focus on adaptation measures for responding to the climate change threat. A cursory review of much of the adaptation literature shows, however, that most of it contains little more than generic recommendations to vaguely defined or specified threats. This situation results from the fact that generally the specific impacts of climate change in specific locations or to specific sectors remain poorly understood and under-investigated. This situation is especially acute in Africa where adaptation strategies are urgently needed. In this chapter, we present the results of a study designed to identify and quantify specific impacts of climate change to the hydro-energy, environmental, and agro-economic sectors in the Niger Basin in West Africa.

The Niger River is the second largest river in Africa by discharge volume (5,600 m^3/s at Onitsha; 1955-1991) and the third longest (4,100 km). Taking its source on the interior side of the Guinea Daro Massif at an altitude of 1068 meters, the Niger River runs off in a northeasterly direction towards the Niger Inland Delta. At Tossaye (Taoussa, Republic of Mali), the river turns southeast forming the Great Bend and flows on through Niger to Lokoja in Nigeria, where it receives its largest and most important tributary, the Benue River (Fig. 1). From there, the enlarged Niger runs directly south to empty into the Bight of Benin through a network of distributaries in Nigeria's Terminal Niger Delta. The river's unusual flow path delineates a drainage basin that is boomerang-shaped and bounded approximately by latitudes 5° N and 22° N and by longitudes 11°30′ W and 15° E. The Niger Basin's total drainage area is 2,170,500 km^2 although 35% (770,500 km^2) of the basin lies within the hyper-arid Sahara and contributes no runoff to the Niger River. The Niger Basin cuts across all the major climatic zones of West Africa, including the Guinean or Equatorial forest zone, the Transitional tropical belt, the Sudan Savanna zone, the semi-arid or Sahel savanna belt, and the desert region. Rainfall, temperature, humidity and evaporation all exhibit steep gradients as one moves hinterland from the coast (Fig. 2).

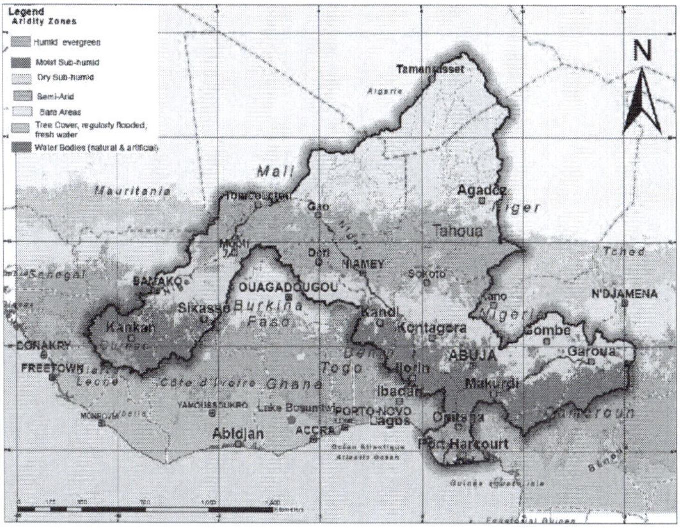

Figure 1. The Location of the Niger Basin in West Africa. Source: BRLi, 2007.

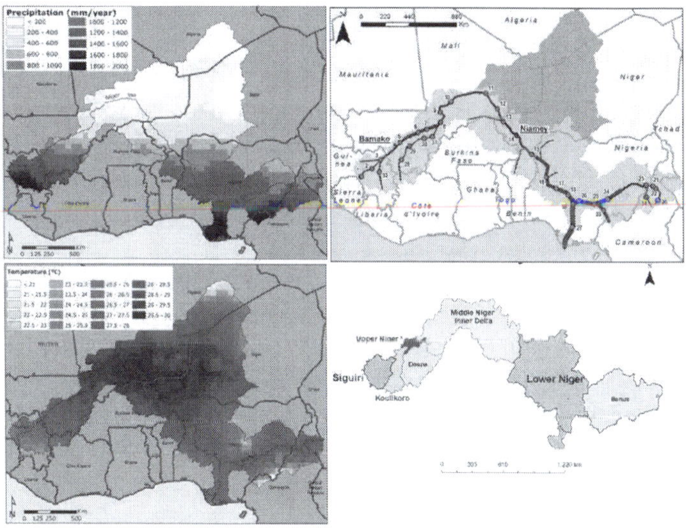

Figure 2. The climatic characteristics of the Niger Basin. The panel on the top left shows the distribution of annual average precipitation, the panel on the top right shows the basin demarcated into active and inactive parts. Also shown is the location of river gauging sites within the basin. The bottom left panel shows the distribution of annual average temperature. Finally, the bottom right panel shows the basin divided into commonly recognized sub-basins. The Upper Niger Basin includes all of the sub basins above the Inner Delta (red area) i.e. Siguiri, Koulikoro (or Sélingué) and Douna (or Bani) watersheds.

The Niger Basin is experiencing considerable pressures on ecosystem resources and with good reason: all 9 countries in the active basin rank among the bottom 40 countries on the Human Development Index (UNDP, 2010). Over 100 million people inhabit the Niger basin distributed in nine countries, namely Benin, Burkina Faso, Cameroon, Chad, Cote D'Ivoire, Guinea, Mali, Niger, and Nigeria (Fig. 1). Average population growth rate is estimated at 2.7% and the rate of urbanization at 4.3%, both numbers among the highest rates in the world (UNFPA, 2010). Approximately 70% of the population depends on agriculture for their livelihood and overwhelmingly on rain-fed agriculture, and 60% lack access to a developed source of water. Somewhat paradoxically, the basins' environmental and water resources are currently grossly underutilized. The irrigation potential is estimated at 2.5 million hectares (ha) but less than 0.3 million ha have been developed; hydropower generation potential is estimated at 30,000 gigawatt hours per year but only 6000 gigawatt hours have been developed; and of the basin's 3,000 km of navigation potential, only 600 km is currently utilized (The World Bank, 2005).

Recognizing the development imperatives, the governments of the member states negotiated and adopted in 2002 a shared vision for development, titled the Sustainable Development Action Plan (SDAP) Under the umbrella of the Niger Basin authority (NBA). With expected total investments of US$ 8 billion over the period 2005 – 2025, the SDAP aims to (i) provide new social and economic opportunities for more than 100 million people living in the Niger Basin, (ii) contribute to capacity building in integrated water resources management, and (iii) protect natural resources and ecosystems (BRLi, 2007). A major part of SDAP (80%) focuses on the construction of three new dams[1] - on the Upper and Middle Niger with associated irrigated agriculture and hydro-energy generation, along with the rehabilitation of existing dams[2] - , and improved navigation and water supply.

As part of SDAP implementation, the Executive Secretariat of the NBA and the World Bank (WB) have jointly implemented a Climate Risk Assessment (CRA) initiative for the NRB. This chapter presents a robust methodology for assessing climate change impacts on the Sustainable Development Action Plan (SDAP).

The CRA initiative is prudent because of a history of marked climate variability with significant socio-economic and environmental impacts. Archival records show evidence of a regional, multi-decadal drought between 1738 and 1756, centered on the Great Bend of the River Niger but extending far south into the coastal areas of present day Ghana. Famine during this period killed half of the population of Timbuktu and other areas of the Niger Bend (Curtin, 1975). In contrast, most of the 1780s experienced excessive rainfall, as far North as Agadez in Niger. More recently, since 1970, the semi-arid Soudan and Sahel savanna zones within the region have witnessed the most dramatic example of inter-decadal climatic variability ever measured quantitatively on the planet since instrumental records have been kept (Rasmusson and Arkin, 1993; Hulme, 2001, p.20; Lebel et al., 2003). Average annual rainfall amounts fell by between 25% and 40%, depending on the location, and persisted over the next two decades, prompting some researchers to hypothesize about whether such decline represented evidence of climate change (Demarée, 1990). Widespread environmental desiccation driven by diminished rainfall and land use practices prompted routine discussions of the "expanding" or "encroaching" Sahara (Lamprey, 1975; Eckholm and Brown, 1977; Los Angeles Times, 1988).

Within this historical context, this CRA is urgent given the expected future impacts of climate change in this region (e.g. IPCC, 2007). Because of the substantial uncertainty in climate projections from the current Global Circulation Models (GCMs), it is difficult to estimate what the future climate is likely to be. This CRA thus aims to better understand the dynamics of the future climate over the NRB, and to assess its potential impacts on water resources, energy, navigation, agriculture and environment, as well as possible consequences for the design of existing and planned infrastructure. Finally, it assesses also the potential negative impacts on the Economic Internal Rate of Return (EIRR) of the SDAP. This is essential for assisting decision makers and stakeholders to better manage their resources, prepare for extreme hydrological hazards, and enhance development planning in the NRB. The CRA assessed the risks of climate change to the water resources and associated development sectors of the Niger Basin in the near (2030), mid (2050), and distant future (2070).

The study is presented in two separate chapters. This chapter introduces the study area, followed by an overview of the study approach and methodology, including the operational definition of risks adopted. Next, the impacts of potential changes in runoff on the SDAP - based on an existing water resources allocation model for the basin - are presented (vulnerability analysis). Finally, the adopted approach to projection of climate changes is described. The next chapter discusses the methodology used in estimating the runoff response to climate change, assess climate risks for key water related sectors, as well as the economic internal rate of return.

OVERVIEW OF APPROACH AND METHODOLOGICAL FRAMEWORK

An essential step for climate related adaptation strategies is to estimate the risk (probability) of exceedance of critical impact levels for various impacted sectors. Such probabilities provide better insights about the likelihood that a specific climate change impact may occur, contributing to informed decisions on investments and adaptations. For this reason, we adopted a risk-based analysis framework. The premise of the study is that estimates of the plausibility of climate risks will help to develop a conceptual framework for adaptation strategies that increase the resilience and robustness of the SDAP investment plan in the NRB.

The adopted methodology is as follows (Ghile et. al, 2013):

i. The impacts of changes in runoff on SDAP performance indicators were estimated using an existing Niger River Basin water resources system model (MIKEBASIN) by parametrically varying the hydrological conditions for the period 1966/67 – 1988/89 (vulnerability analysis). The runoff elasticities of selected performance indicators[3] - were then determined (Brown and Grijsen, 2013).
ii. Based on the Intergovernmental Panel on Climate Change (IPCC) A1B scenario, we estimated the probability distributions of changes in precipitation and temperature, using a bias corrected ensemble of 38 Global Climate Model (GCM) projections of future climate for the 21st century.
iii. Using a variety of hydrologic models and methods, we determined the response of runoff to climate changes (defined as climate elasticity[4] - of runoff) and translated the probability distributions of future changes in precipitation and temperature into probability distributions of future mean runoff (Grijsen and Brown, 2013).
iv. Finally, we estimated the probabilities of changes in SDAP performance based on impacts of runoff changes and the assessed probability distributions of changes in runoff.

Defining Climatic Risk in the Niger Basin

In this study climate risk is defined on the basis of changes in annual runoff as a function of projected precipitation and temperature changes; it does not take into account other sources of risk such as vulnerability of specific sectors and future socio-economic changes, nor does it account for changes in the variability of runoff within the year. As mentioned previously, we used an ensemble of 38 climate model runs to account for model error and natural climate variability, and also to obtain the full range of possible climate futures and associated runoff. Each run was treated as equally plausible. Using these runs the probability distributions of future runoff regimes, centered on the years 2030, 2050 and 2070 were fitted to estimate the plausibility of climate risks. To assess the severity and probability of climate-imposed risks in the basin, the generated future mean flows from the 38 climate projections were used to estimate risk levels for key performance metrics. In consultation with basin stakeholders in a CRA workshop in Ouagadougou (2010), reductions of more than 20% from the baseline performance were considered to be significant impacts. With this guidance, we identified five risk levels ranging from "Mild risk" to "Extreme risk" (Fig. 3), based on a choice of percentage shifts in performance relative to baseline operations. In this study the baseline refers to a development scenario in which three dams are built, along with associated irrigation infrastructure. These dams include Fomi (FO; in the Republic of Guinea), Taoussa (TA; in the Republic of Mali) and Kandadji (KD; in the Republic of Niger). Thus, the baseline scenario is interchangeably referred to as the FO-TA-KD scenario. We focused on five sectors and key performance metrics, as identified by the Ouagadougou CRA workshop participants and shown in Table 1.

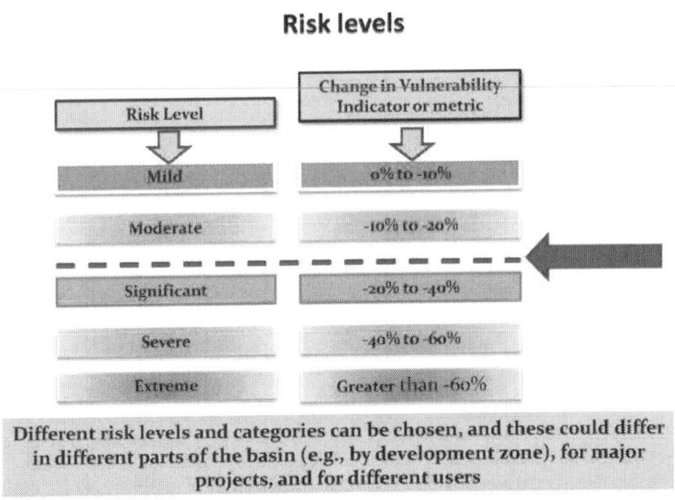

Figure 3. Risk levels based on change in vulnerability indicators. The red arrow indicates the 20% reduction in performance relative to base level beyond which impacts are considered significant.

Table 1. Key performance indicators and baseline values for various sectors

Sector/ Domain	Indicator	Baseline
Agriculture/irrigation	Incremental net irrigated area	1.5 Mha
Energy production	Total annual energy production	8,250 GWh
Navigation	Annual number of days	5 months
Inner Delta Flooding	Average annual flooded area	11,000 km^2
Environmental Flows (at Markala, Mali-Niger border, Niamey and Malanville)	Minimum flow in m^3/s	40, 75, 125 and 80 m^3/s

Climate Projections

Climate projections typically lack credibility at the spatial and temporal scales that are relevant to water resources planning, especially in tropical regions where monsoonal systems are poorly represented and inter-annual variability is poorly reproduced. While a variety of downscaling approaches are available which change the resolution of climate projections, they do little to improve their credibility. In this analysis we have attempted to address these concerns in two ways. First, as described above, a bottom-up (decision-scaling) approach to assessing risks was adopted (Brown et. al, 2012), which focuses primarily on the vulnerability of the water resources system performance to runoff changes due to climate change. As a result, the climate projections are placed in the context of risks to investments rather than as credible projections of the future. Second, the climate projections were not downscaled to higher resolution; instead, the hydrologic response of the basin was "up-scaled" to better match the resolution at which climate projections have more credibility, i.e. at the basin scale. These approaches represent an attempt to make the best use of the highly uncertain climate information described further below.

Precipitation projections for West Africa vary widely, such that GCMs even lack agreement on the direction of future changes in precipitation (Figs. 4 - 5). Therefore, we analyzed precipitation and temperature projections from an ensemble of 38 GCM simulations[5] - for the 21st century, to capture a quantitative global assessment of climate-change-driven risks for the key metrics identified in the NRB. The projections were driven by the median A1B CO_2 emission scenario (Nakicenovic and Swart, 2000). To correct for systematic biases inherent in all GCM models, we employed the quantile mapping method - as described in Wood (2002), Wood (2004) and Segui (2010), and trained it on observed and simulated data for the 20th century. It was initially assessed that the estimation of climate risks over the entire NRB would be the most skillful scale for projections. This analysis was done for the periods 2016-2045, 2036-2065 and 2056-2085, considered representative respectively for 2030, 2050 and 2070.

We have also used climate change projections available for Guinea and Nigeria from the climate wizard website[6] - (climatewizard.org) and the WBG's Climate Change Knowledge Portal (Climate Portal)- as being representative for respectively the Upper Niger Basin and the Lower Niger and Benue Basins - to determine whether climate change impacts in the Upper Niger (Guinea) and Benue (mainly Nigeria) Basins could be more or less severe than the average impacts across the Niger Basin. The Upper Niger and Benue Basins constitute the main water sources ('water towers') of the Niger River. Results are shown in Table 2, showing a good agreement between the various sources of climate change projections.

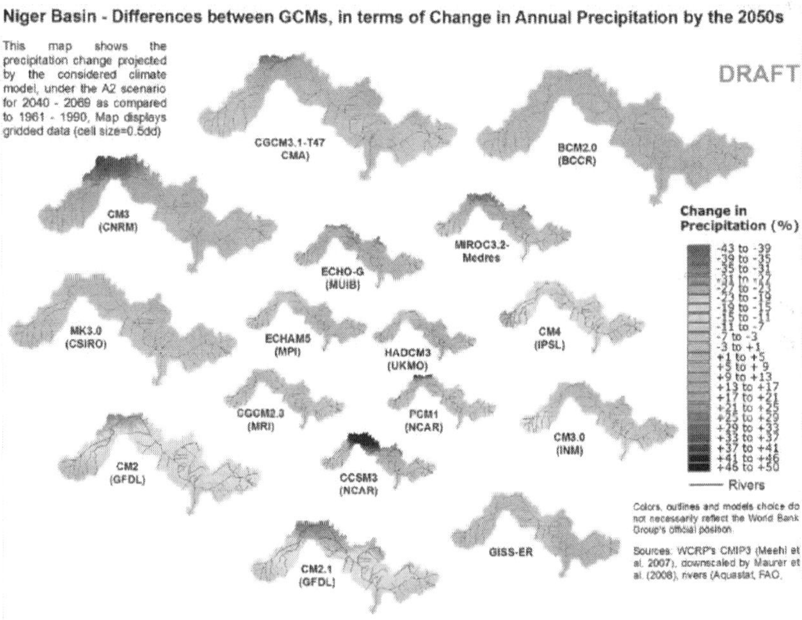

Figure 4. Annual precipitation changes projected for 2050 by various GCMs[7] - (Sources: right lower corner)

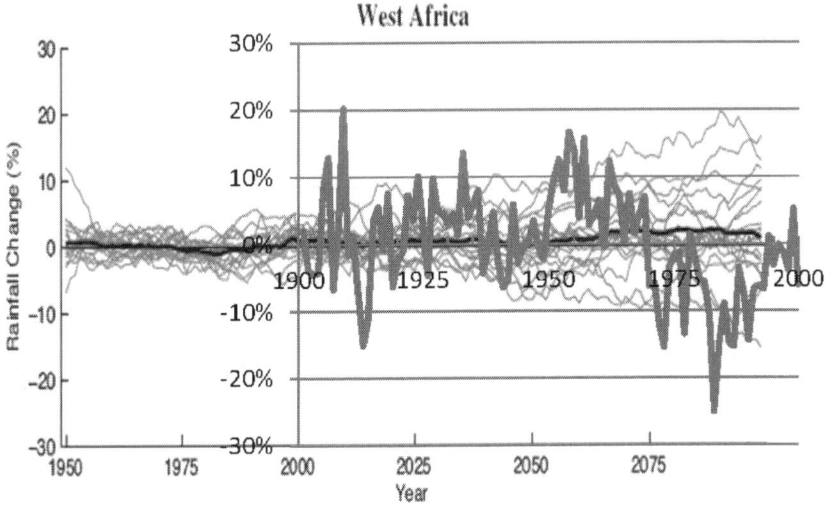

Figure 5. Ensemble of regional climate projections for West Africa (Giannini et al., 2008); the black line is the ensemble mean (A1B); the bold blue line reflects the historical 20th century rainfall variability for the Niger River Basin

Strzepek (2011) provides a detailed account of the studies and hydrological modeling underpinning the results displayed in the Climate Portal. To generate a high-resolution understanding of possible risks, the analyses examined relative changes in several hydrological indicators (including inter alia annual precipitation, temperature and mean runoff, high flow, low flow and potential evapotranspiration (E_0)) from the historical baseline (1961 to 1999) to the 2030s and 2050s, for a range of 56 GCM and emission scenario combinations. Projected changes in runoff and E_0 were simulated with the CLIRUN-II hydrologic model II (Strzepek and Fant, 2010), an extension of the WatBal water balance model (Yates, 1996). The results provide an understanding of the range of potential consequences of climate change on water resources at the country and basin scale. The results are claimed to be suitable for use as inputs to screening-level analyses of the impact of climate change on water resources dependent investments, for the 2050 investment horizon. Results for the various sources of climate projections agree very well with the results obtained under the present NRB-CRA study on the basis of 38 GCM model runs. Differences between impacts for the two countries are marginal. The Climate Portal shows potential evapotranspiration to increase by nearly 5% in 2050, or a climate sensitivity of E_0 of 2.3% per ^{0}C at an average annual basin temperature of 26.6 ^{0}C.

Table 2. Summary of projected climate changes from various sources (2050; A1B)

Variable	Min	20%	Mean	80%	Max	St. dev.
Climatewizard.org						
Guinea						
Temperature (^{0}C)	1.8	2.0	2.3	2.8	3.0	
Precipitation (%)	-20.0	-6.0	0.0	6.0	10.0	
Nigeria						
Temperature (^{0}C)	1.5	1.8	2.1	2.5	2.8	
Precipitation (%)	-15.0	-4.0	2.0	10.0	15.0	
WB Climate Change Knowledge Portal						
Guinea						
Temperature (^{0}C)	1.2	1.8	2.1	2.6	3.0	0.5
Precipitation (%)	-12.2	-5.2	0.5	5.6	12.9	6.8
Annual runoff (%)	-23.8	-13.5	-0.3	12.0	38.7	16.5
Annual PET (%)	0.7	3.9	5.0	6.7	8.1	1.7
Nigeria						
Temperature (^{0}C)	1.2	1.6	2.0	2.4	2.7	0.4
Precipitation (%)	-13.4	-4.4	1.2	7.0	10.9	6.4
Annual runoff (%)	-31.0	-11.3	-0.2	15.1	29.9	17.0
Annual PET (%)	1.5	3.7	4.6	6.1	7.4	1.4
Projections 38 GCM model runs for Niger River Basin						
Temperature (^{0}C)	1.2	1.6	2.1	2.6	2.9	0.5
Precipitation (%)	-5.8	-3.5	1.4	4.5	13.7	4.5
Annual runoff (%)	-19.5	-13.2	-1.9	4.7	32.3	10.9
Annual PET (%)	2.6	3.6	4.7	5.8	6.7	1.1

The results show that a slightly larger number of models project increases in precipitation than those projecting decreases, while all the models project increases in temperature. Fig. 6 shows quartiles of rainfall and temperature projections (0, 25, 50, 75 and 100% points), for percentage changes relative to the 20th century averages. The graphs show variations of the 30-yr average across the 38 GCM model runs, for 2030, 2050 and 2070, interpreted as the variance of the projected future mean runoff.

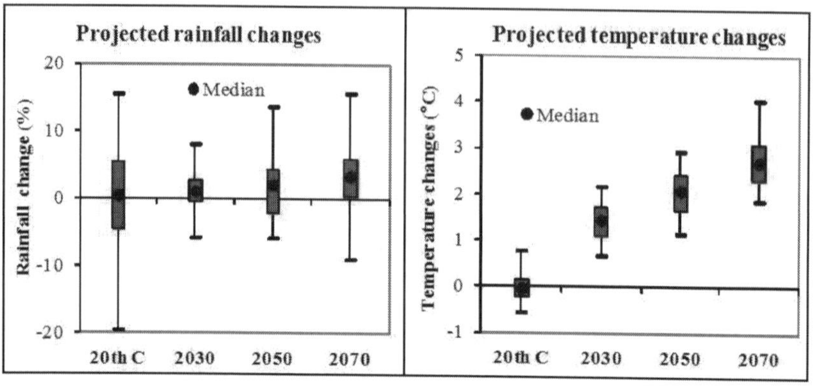

Figure 6. Quantiles for rainfall and temperature changes[8] - for the 20th century and projections for 2030, 2050 and 2070

Fig. 7 depicts annual rainfall and temperature projections for the 21st century averaged for each year over all GCM runs. It shows hardly any upward trend in precipitation till 2050 and 3% by 2070; 34 (90%) of the 38 model runs project by 2050 changes between -6% and +7%. Temperatures show a steady increase of 0.03^{0}C per year (2.9 ^{0}C over 100 years); on average 2.1^{0}C (8%) by 2050 and 2.7^{0}C by 2070 (10%). Projected temperature increases vary between +1.0^{0}C and +3.0^{0}C by 2050. The primary impact of increased temperatures by 2050 is an increase of the potential evapotranspiration by about 5% for a 2.1^{0}C increase in temperature (Grijsen, 2013), and a similar increase in gross irrigation requirements. Catchment runoff is generally also reduced due to increased evaporation, as discussed in Part II. The normal distribution has been shown to apply well to the 30-year average projections of precipitation and temperature for 2050 (Fig.8).

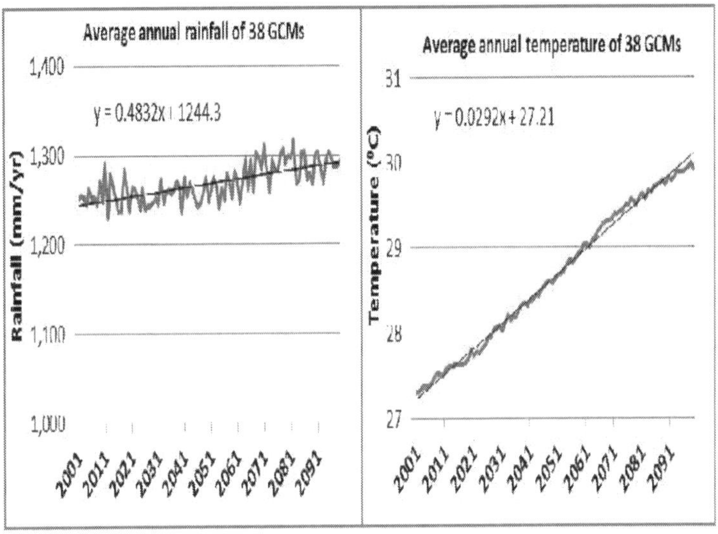

Figure 7. Annual rainfall and temperature projections for the 21st century averaged over all GCMs runs

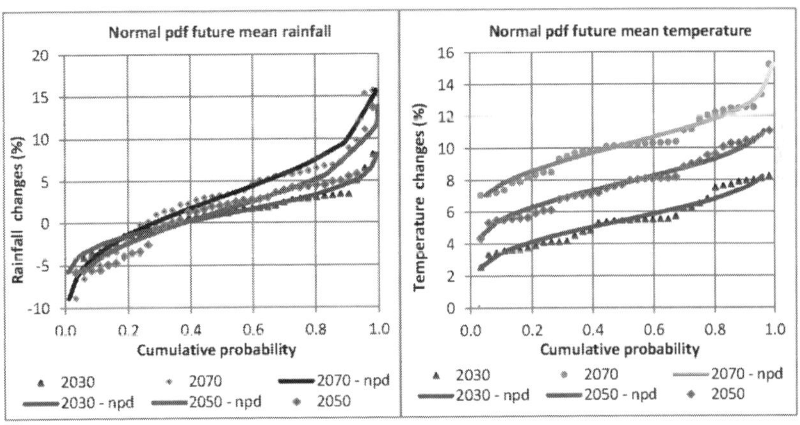

Figure 8. Normal distribution of projected average precipitation and temperature for 2030 – 2070

Assessing Impacts of Potential Runoff Changes on Key Sdap Performance Indicators

In 2007, a Mike-Basin water resources system model was developed for the NRB to study the impact of various SDAP development scenarios on the water resources of the Basin (BRLi, 2007). We have adopted this model to simulate the water availability, water demands and water abstractions from the river and reservoir system within the NRB. The Mike Basin model is designed to simulate and assess the balance between the abstractions of water for irrigation and domestic and industrial water supply on the one hand, and energy production, flooding, navigation and environmental constraints (required minimum flows) at any location of interest on the other hand. The model requires input in the form of hydrological data (including streamflow, rainfall and potential evapotranspiration), reservoir characteristics and operating rules, minimum environmental flows, water demand series for domestic needs and irrigation, hydro-electric power generation parameters, and multiple socio-economic parameters (for irrigated agriculture, hydro-energy, navigation, environmental benefits and cost, etc.). The NRB has been divided into 60 sub-catchments, with 420 river nodes, 21 existing and 4 planned dams, 10 hydropower plants, and 93 water abstraction points for various uses (Fig. 9). The model was calibrated and verified with historical hydro-meteorological data for the period 1966/67 – 1988/89. This period contains some years with above average flow conditions during the 1960s, but is dominated by the drought conditions during the 1970s and 1980s, with below average flow conditions as shown inFig. 10.

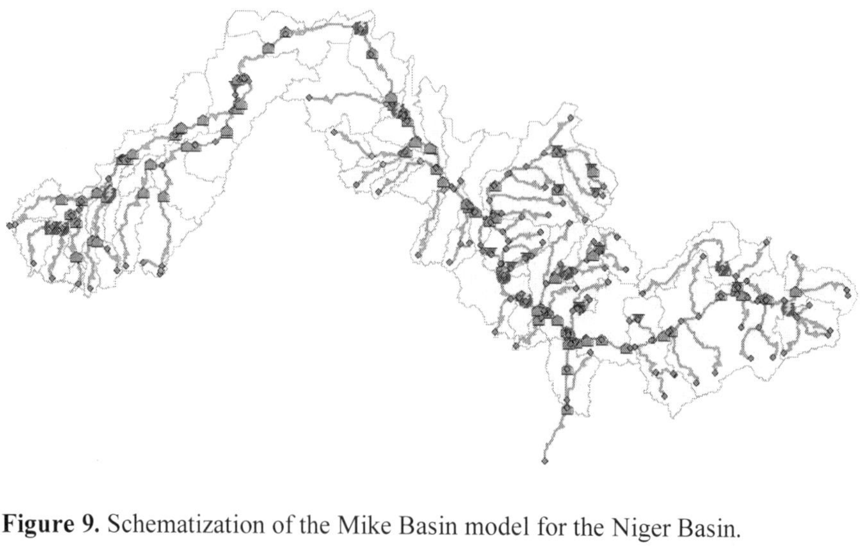

Figure 9. Schematization of the Mike Basin model for the Niger Basin.

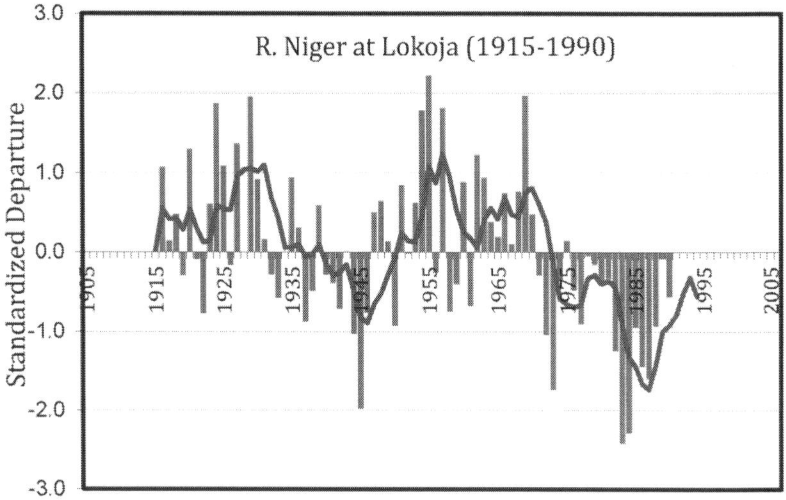

Figure 10. Standardized annual flow data for Lokoja at the confluence of the main Niger and Benue Rivers.

The relationships between relative changes in runoff and relative changes in specific performance criteria at basin level were derived from numerous Mike-Basin runs for the FO-TA-KD development scenario, for parametrically induced variations in runoff between -30% and + 10% of the 20th century baseline conditions, and water demands with a 5% increase compared to the 20th century baseline (Table 3). The average projected temperature increase by 2050 is 2.1^{0}C. Based on the Penman-Monteith and modified Hargreaves methods this is estimated to cause an increase of 5% in potential evapotranspiration and crop water demand (Grijsen, 2013). Fig. 11 shows the runoff changes required to cause specific changes in selected performance indicators, such as -10%, -20% etc. The 1/5 yrs (20%) performance indicators are slightly more sensitive to runoff changes than the average performances.

Under the FO-TA-KD development scenario irrigated agriculture will receive a major boost, increasing the average total harvested area by 440%. The average basin wide hydro-energy production is estimated to increase due to SDAP by 13% (including the contributions of multiple run-of-river hydro-plants) and the average navigable months in the NRB are estimated to decrease by about 20%. Flooding of the Inner Delta will face a decrease of about 10% due to SDAP. However, hydro-energy production at Kainji and Jebba dams on the main stem Niger River in Nigeria will face a decrease of nearly 35%, caused by the large incremental upstream irrigation water withdrawals. Minimum flows in the Middle Niger will be significantly improved, while minimum flows through the Inner Delta will remain well above the required 40m^3/s.

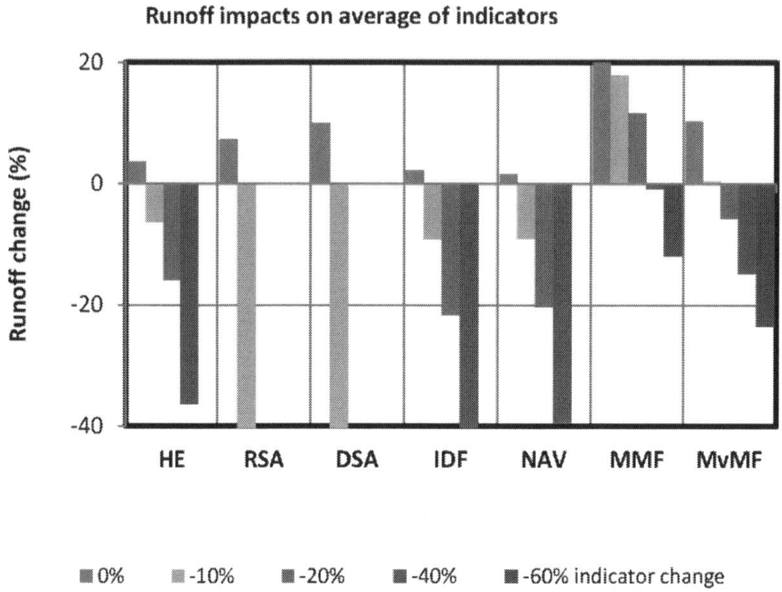

Figure 11. Runoff changes required to generate specific changes in performance indicators

Table 3 demonstrates the sensitivity of performance indicators for changes in runoff. Significant reductions of more than 20% of the baseline performance generally occur for hydro-energy production, flooding of the Inner Delta and navigation when the long-term average basin runoff decreases by 15 to 20%, or more. Such long-term average decrease would represent a major increase in the aridity of the basin[9] - . This sensitivity to runoff changes is attributed to the fact that energy production is not prioritized in the applicable water allocation rules under the existing Water Charter for the Niger Basin. The Charter assigns the highest priority to irrigation and domestic water supply to secure food production and alleviate poverty. Thus, irrigated agricultural is the least sensitive to climate change impacts on basin runoff. Moreover, during the rainy season water availability remains abundant and secures in most years full replenishment of reservoirs for dry season irrigation, even after significant flow reductions. Minimum flows at Markala (upstream of the Inner Delta) and in the Middle Niger (Table 3) are extremely sensitive to changes in basin runoff and cannot be maintained even under a 10% reduction in runoff. This is reflected in the runoff elasticity of performance indicators (last column ofTable 3), defined as the ratio between relative changes in an indicator and relative changes in runoff[10] - . Runoff elasticities of performance indicators can thus be divided in three groups, i.e. (i) irrigated agriculture with a runoff elasticity of 0.1 to 0.2 (non-sensitive); (ii) hydro-energy, navigation and Inland Delta flooding with runoff elasticities of 1.0 to 1.3 (sensitive), and (iii) minimum flows with elasticities of 2 to 4 (extremely sensitive). The dry season minimum flows are also very sensitive to increased dry season irrigation demands due to climate change.

Table 3. Performance indicator matrix for the FO-TA-KD scenario (-10% R means a 10% reduction of runoff)

Performance Metrics	Probability of non-exceedance	Reference 2005 (SA)	Impacts FO-TA-KD	20^{th}C + 5% water demands; FO-TA-KD						Average runoff elasticity
				20^{th}C-R	+10% R	0% R	-10% R	-20% R	-30% R	
		Value	% change	Value	Percentage changes (%)					
Hydro-energy										
Basin energy	1/2(50%)	7,376	12.6	8,303	5.2	-3.0	-13.9	-24.3	-33.8	**1.0**
(GWh)	1/5(20%)	5,771	3.4	5,969	8.9	-3.5	-16.0	-28.5	-38.7	**1.3**
Kainji/Jebba	1/2(50%)	5,007	-32.5	3,381	11.1	-5.8	-23.0	-37.3	-48.2	**1.7**
	1/5(20%)	3,743	-43.9	2,100	9.8	-5.1	-21.3	-38.0	-50.3	**1.7**
Irrigated Agriculture										
Total irrigation	mean	228,138	435	1,220,591	0.1	-0.3	-0.9	-1.8	-3.6	**0.1**
RS (ha)	1/5(20%)	228,138	424	1,194,537	-0.1	-0.5	-2.1	-4.2	-8.1	**0.2**
Total irrigation	mean	111,744	471	637,537	-0.4	-0.8	-1.2	-1.5	-2.2	**0.1**
DS (ha)	1/5(20%)	105,130	500	630,890	-0.7	-0.9	-1.4	-5.5	-15.7	**0.3**
Navigation for various reaches (average number of days)										
Average	Large boats	171	-20.9	135	7.9	-1.4	-10.9	-19.7	-30.2	**1.0**
Flooding (km^2)	mean	12,117	-9.7	10,940	5.4	-1.5	-10.9	-18.7	-28.6	**0.9**
Inland Delta	1/5(20%)	10,342	-14.1	8,887	7.1	-1.6	-13.9	-24.8	-37.3	**1.2**
Sustenance of 10-day average minimum flows (m^3/s)										
Markala	1/2(50%)	70	-13	61	-25.0	-38.6	-54.7	-83.5	-100.0	**4**
	1/5(20%)	51	-2	50	-31.1	-40.6	-69.8	-99.6	-100.0	**5**
Mali-Niger	1/2(50%)	60	33	80	-2.7	-12.3	-32.6	-65.2	-89.0	**3**
border	1/5(20%)	12	547	80	-15.9	-32.3	-63.9	-82.1	-97.4	**4**
Niamey	1/2(50%)	55	78	99	2.0	-9.2	-16.9	-37.5	-80.9	**2**
	1/5(20%)	9	818	85	-0.9	-11.3	-33.6	-73.7	-97.0	**3.5**
Malanville	1/2(50%)	68	35	91	-0.3	-10.3	-27.2	-53.9	-71.6	**2.5**
	1/5(20%)	4	2,035	77	-2.1	-21.9	-55.8	-83.5	-98.2	**4**

CONCLUSIONS

Effectively managing climate change will constitute one of the major challenges of the 21st. This challenge will be especially acute in those regions of the world already prone to adverse climatic variability and where the financial and technical resources for dealing with climate change are lacking. The River Niger Basin in West Africa fits this profile. However, in collaboration with several donor and investment partners, the member states of the Niger Basin under the aegis of the Niger Basin Authority are implementing a novel climate risk assessment study in an effort to climate-proof new investments. This chapter described a part of that study, including an over view of the methodology, a description of the climate change estimation procedure and an assessment of the impact of potential runoff changes on key SDAP performance indicators. The key conclusions of the study will be summarized in the next chapter (part II).

Climate change confronts decision makers with deep uncertainty, where decisions can go wrong if decision makers assume that risks are well-characterized while in reality uncertainties could be underestimated. Such conditions require robust decision making to inform good decisions by identifying system vulnerabilities and assessing alternatives for ameliorating those vulnerabilities. A water resources system model and the Performance Indicator Matrix (Table 3) prepared on the basis of a large number of model simulations, provide a powerful tool to identify early on in the CRA process, even before climate projections have been processed, system vulnerabilities and future climate change conditions where the proposed SDAP plan may fail to meet its goals. These conditions can then be used to early identify potential actions to address vulnerabilities and evaluate tradeoffs among them. This approach, which focuses on water resources system vulnerability rather than on climate change projections, is considered crucial in view of the large uncertainties embedded in the presently available climate projections for the 21st century.

Notes

[1] - Fomi dam in Guinea, Taoussa dam in Mali and Kandadji dam in Niger, referred hereafter as the FO-TA-KD development scenario.

[2] - Kainji and Jebba dams on the main Niger River in Nigeria, and Lagdo dam on the Benue River in Cameroon.

[3] - The runoff elasticity of a performance indicator (e.g. hydro-energy production) defines the response (as a multiplier) of the indicator to changes in runoff; for example, a runoff elasticity of hydro-energy of 1.2 indicates that a 10% decrease in runoff would cause a 12% decrease in generated hydro-energy.

[4] - Climate elasticity of runoff defines the response (as a multiplier) of runoff to changes in precipitation and temperature; for example, a

precipitation elasticity of runoff of 2.5 indicates that a 10% decrease in rainfall would cause a 25% decrease in runoff. Similarly, a temperature elasticity of runoff of -0.5 indicates that an increase in temperature of 10% (e.g. from 26 to 28.60C) would cause a decrease in runoff of 5%.

[5] - Source: Data library of the International Research Institute for Climate and Society (IRI).

[6] - University of Washington and the Nature Conservancy (2009); Data source: Global Climate Model (GCM) output, from the World Climate Research Program's (WCRP) Coupled Model Inter-comparison Project phase 3 (CMIP3) multi-model dataset (Meehl et al., 2007), was downscaled (as per Maurer et al., 2009) using the bias-correction/spatial downscaling method of Wood et al. (2004) to a 0.5 degree grid, based on the 1950-1999 gridded observations of Adam and Lettenmaier (2003).

[7] - The largest changes are projected for regions with low rainfall, which has small implications in absolute terms.

[8] - The average baseline temperature for the 20th century is 26.60C
[9] - It is noted that the droughts of the 1970s and 1980s are already reflected in the 2005 baseline flow regime.

[10] - A runoff elasticity of +1.5 indicates that a 10% decrease in runoff causes a 15% decrease in performance.

ACKNOWLEDGEMENTS

This study was done as part of the Climate Risk Assessment (CRA) for the Niger Basin, a joint initiative of the Niger Basin Authority (NBA) and the World Bank to assess the risks from climate change to the performance of NBA's Sustainable Development Action Plan (SDAP) for the Niger Basin. The aim of this initiative is to build resilience to climate risks into the SDAP. The Niger CRA study is supported by grants from (i) the Bank-Netherlands Partnership Program (BNPP) Trust Fund, (ii) the Trust Fund for Environmentally and Socially Sustainable Development (TFESSD) funded by Finland and Norway, (iii) the Norwegian Trust Fund (NTF) and (iv) the Trust Fund for Integrated Land and Water Management for Adaptation to Climate Variability and Change (ILWAC) funded by Denmark. We gratefully acknowledge the NBA Observatory for making hydro-meteorological and runoff data available, as well as providing valuable comments and directions for the study. We extend our thanks to Dr. Amal Talbi for managing this CRA study for the World Bank, and to the World Bank's Water Partnership Program (WPP) and Water Unit for organizing the special session S3 of HydroPredict2012 (Vienna, September 2012) on 'Choosing Models for Resilient Water Resources Management' (Grijsen et. al, 2013), and for giving permission to present the results of this CRA work in this book. We thank the World Bank for permission to report this work here.

REFERENCES

1. BRLi et alJanuary 2007Evaluation des prélèvements et des besoins en eau pour le modèle de simulation du basin du Niger, Final Report.
2. BRLi et alJuly 2007Elaboration du Plan d'Action de Développement Durable du Bassin du Niger, Phase II : Schéma Directeur d'Aménagement et de Gestion.
3. BRLi et alSeptember 2007Establishment of a Water Management Model for the Niger River Basin, Final Report.
4. C Brown, and J. G Grijsen, 2013Sensitivity of SDAP performance to changing water availability and demands; the Niger River Basin Climate Risk Assessment, a joint initiative of the Niger Basin Authority and the World Bank (not yet disclosed).
5. C Brown, Y Ghile, M Laverty, and K Li, 2012Decision scaling: linking bottom-up vulnerability analysis with climate projections in the water sector,Water Resources Research, 48(9), 1-12
6. P. D Curtin, 1975Economic Change in Pre-Colonial Africa: Supplementary Evidence, University of Wisconsin Press, Madison.
7. G. R Demaréee, 1990An indication of climatic change as seen from the rainfall data of a Mauritanian station, Theoretical and Applied Climatology, 42139147
8. E Eckholm, and L. R Brown, 1977Spreading deserts-The hand of man, World watch Paper 13Washington D.C.
9. A Giannini, M Biasutti, I. M Held, and A. H Sobel, 2008A global perspective on African climate, Climatic Change 904359383
10. Y. B Ghile, M. Ü Taner, C Brown, J. G Grijsen, 2013Bottom-up Climate Risk Assessment of Infrastructure Investment in the Niger River Basin, Climate Change (under review)

11. J. G Grijsen, 2013Future Water Demands in the Niger Basin, the Niger River Basin Climate Risk Assessment, a joint initiative of the Niger Basin Authority (NBA) and the World Bank (not yet disclosed).
12. J. G Grijsen, and C Brown, 2013Climate elasticity of runoff and climate change impacts in the Niger River Basin; the Niger River Basin Climate Risk Assessment, a joint initiative of the Niger Basin Authority (NBA) and the World Bank (not yet disclosed).
13. J. G Grijsen, C Brown, A Tarhule, Climate Risk Assessment for Water Resources Development in the Niger River Basin; HydroPredict'20122012International Conference on Predictions for Hydrology, Ecology and WRM, Special Session S3- Choosing models for resilient water resources management, Vienna; World Bank WET/WPP/TWIWA publication (in print).
14. M Hulme, 2001Climatic perspectives on Sahelian desiccation: 1973-1998. Global Environmental Change 111929
15. IPCCKundzewicz, Z.W., L.J. Mata, N.W. Arnell, P. Döll, P. Kabat, B. Jiménez, K.A. Miller, T. Oki, Z. Sen and I.A. Shiklomanov, 2007Freshwater resources and their management. Climate Change 2007: Impacts, Adaptation and Vulnerability. Contribution of Working Group II (Chapter 3) to the Fourth Assessment Report of the Intergovernmental Panel on Climate Change, M.L. Parry, O.F. Canziani, J.P. Palutikof, P.J. van der Linden and C.E. Hanson, Eds., Cambridge University Press, Cambridge, UK, 173210
16. H. F Lamprey, 1975Report on the desert encroachment reconnaissance in northern Sudan, 21 Oct. to 10 Nov., 1975; UNESCO/UNEP.
17. T Lebel, J. L Redelsperger, and C Thorncroft, 2003African Monsoon Multidisciplinary Analysis (AMMA): an international research project and field campaign; CLIVAR Exchanges 85254
18. Los Angeles Times1988Desert Encroachment: Fabled Oasis of Timbuktu Is Drying Up; byline by Scott Kraft, March 08, 1988; available online athttp://articles.latimes.com/1988-03-08/news/mn-608_1_desert-encroachment;last accessed September 3, 2012.
19. E. P Maurer, J. C Adam, and A. W Wood, 2009Climate Model based consensus on the hydrologic impacts of climate change to the Rio Lempa basin of Central America, Hydrology and Earth System Sciences13183194
20. G. A Meehl, C Covey, T Delworth, M Latif, B Mcavaney, J. F. B Mitchell, R. J Stouffer, and K. E Taylor, 2007The WCRP CMIP3 multi-model dataset: A new era in climate change research, Bulletin of the American Meteorological Society, 8813831394
21. Nakicenovic and Swart2000Special Report on Emissions Scenarios, Cambridge University Press, Cambridge, UK. Available athttp://www.grida.no/climate/ipcc/emission/023.htm.

22. E. M Rasmusson, and P. A Arkin, 1993A global view of large-scale precipitation variability. Journal of Climate 614951522
23. P. Q Segui, A Ribes, E Martin, F Habtes, and J Boe, 2010Comparison of three downscaling methods in simulating the impact of climate change on the hydrology of Mediterranean basins, Journal of Hydrology 383111124
24. K. M Strzepek, A Mccluskey, B Boehlert, M Jacobsen, and C. W Fant, IV, 2011Climate Variability and Change: A Basin Scale Indicator Approach to Understanding the Risk to Water Resources Development and Management, Water Anchor of the World Bank Group, series Water Papers: WB Water Paper CCK Portal.
25. K. M Strzepek, and C. W Fant, I. V Water, and Climate Change: Modeling the Impact of Climate Change on Hydrology and Water Availability, University of Colorado and Massachusetts Institute of Technology.
26. United Nations Development Program2010Human Development Indicators-2010 Rankings. The United Nations Development Program. Available online at http://hdr.undp.org/en/media/PR3HDR10-HD1-E-rev4.pdf.
27. United Nations Population Fund2010The state of the world population: Demographic and economic indicators. UNPFA. Available online athttp://www.unfpa.org/swp.
28. World Bank2005The Niger River Basin: A vision for sustianable management. The World Bank. Available online athttp://elibrary.worldbank.org/content/book/9780821362037.
29. D Yates, 1996WatBal: An integrated water balance model for climate impact assessment of river basin runoff. International Journal of Water Resources Development 122121139
30. A. W Wood, E. P Maurer, A Kumar, and D. P Lettenmaier, 2002Long-range experimental hydrologic forecasting for the eastern United States, Journal of Geophysics Research 107(D20), 4429.
31. A. W Wood, L. R Leung, V Sridhar, and D. P Lettenmaier, 2004Hydrologic implications of dynamical and statistical approaches to downscaling climate model outputs, Climate Change 62189216

CITATION

J. G. Grijsen, C. Brown, A. Tarhule, Y. B. Ghile, Ü. Taner, A. Talbi-Jordan, H. N. Doffou, A. Guero, R. Y. Dessouassi, S. Kone, B.

Coulibaly and N. Harshadeep (2013). Climate Risk Assessment for Water Resources Development in the Niger River Basin Part I: Context and Climate Projections, Climate Variability - Regional and Thematic Patterns, Dr. Aondover Tarhule (Ed.), ISBN: 978-953-51-1187-0, InTech, DOI: 10.5772/56707. Available from: http://www.intechopen.com/books/climate-variability-regional-and-thematic-patterns/climate-risk-assessment-for-water-resources-development-in-the-niger-river-basin-part-i-context-and-

CHAPTER 8

Climate Risk Assessment for Water Resources Development in the Niger River Basin Part II: Runoff Elasticity and Probabilistic Analysis

J.G. Grijsen[1], A. Tarhule[2], C. Brown[3], Y.B. Ghile[4], Ü. Taner[5], A. Talbi-Jordan[6], H. N. Doffou[7], A. Guero[7], R. Y. Dessouassi[7], S. Kone[7], B. Coulibaly[7] and N. Harshadeep[8]

[1]Independent Hydrology and IWRM Consultant, Arrowlake Road, Wimberley, Texas, USA
[2]Department of Geography and Environmental Sustainability, University of Oklahoma, Norman, USA
[3]Department of Civil and Environmental Engineering, University of Massachusetts, Amherst, USA
[4]Woods Institute for the Environment, Stanford University, Stanford, USA
[5]Dept. of Civil and Environmental Engineering, University of Massachusetts, Amherst, USA
[6]The World Bank Middle East North Africa (MNSWA), Washington, USA
[7]Niger Basin Authority/ Autorité du Bassin du Niger (ABN), Niamey, Niger
[8]The World Bank, Africa Region, Washington DC, USA

INTRODUCTION

The preceding chapter presented the context and overview of the climate risk assessment (CRA) for the Niger River Basin (NRB). It also discussed the climate change estimation methodology and the impact of potential runoff changes on the performance indicators of the Niger Basin

Sustainable Development Action Plan (SDAP). In this chapter, we describe the methodology used in estimating the runoff response to climate change, followed by the quantitative assessment of climate risks for key water related sectors, and a discussion of the major findings.

RUNOFF RESPONSE TO CLIMATE CHANGE

For this study, the primary goal is to determine the relative changes (in %) in annual runoff and various performance criteria such as hydro-energy and irrigated agriculture production, due to relative changes (in %) in annual climate parameters (notably the precipitation, P, and temperature, T) caused by projected future climate changes. The task, then, is to determine the response of runoff to changes in climate parameters, i.e. to estimate the climate elasticity of runoff. A precipitation elasticity of runoff of, for example, 2.5 implies that a 10% increase in precipitation causes a 25% increase in runoff; a temperature elasticity of -0.5 implies that a 10% increase in temperature causes a 5% decrease in runoff. For this purpose we have extensively reviewed available literature on climate elasticity of basin runoff,inter alia Wigley and Jones (1985), Gleick (1986, 1987), Karl and Riebsame (1989), Risbey and Entekhabi (1996), Vogel et al (1999), Arora (2002), Legates et al (2005), including the estimation of climate change impacts on runoff through hydrological modeling studies for similar river basins in Africa, inter alia Deksyos Tarekegn (2006), Strzepek and McCluskey (2006) and SNC Lavalin (2007). We also applied linear and log-linear regression models to the available rainfall, temperature and runoff data for various sub-catchments of the Niger Basin (Grijsen and Brown, 2013). The data used were spatially aggregated (gridded) annual precipitation and temperature data for the period 1948-2002 for major sub-catchments of the Niger Basin (Hirabayashi, 2008) and annual stream flow data from the data base of the Niger Basin Observatory of the Niger Basin Authority (NBA). The Basin's network of hydrometric stations is shown in Fig. 1. In the preceding chapter, we showed that the projected changes in annual precipitation and temperature were well represented by the normal distribution.

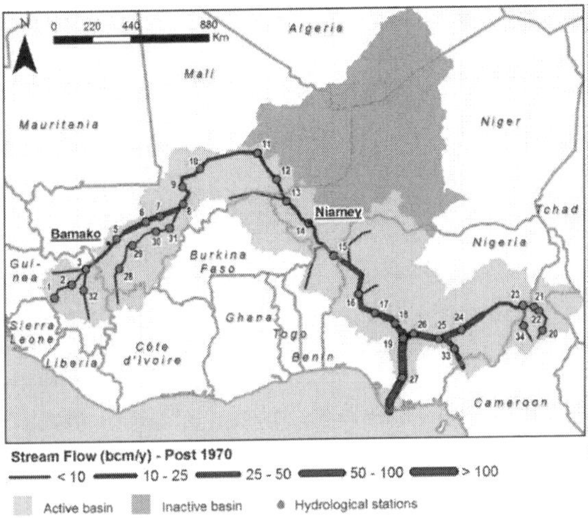

#	Station Name	Pre-1970 Flow	Post-1970 Flow
1	Faranah	2.5	1.7
2	Kouroussa	8.1	5.2
3	Siguiri	39.0	23.6
5	Koulikouro	48.9	32.9
6	Kirango aval	50.9	26.1
7	Ké-Macina	50.6	30.4
8	Mopti	42.9	23.3
9	Akka	-	24.5
10	Diré	36.6	24.5
11	Taoussa (Tossaye)	-	20.5
12	Ansongo	34.3	20.4
13	Kandadji	-	22.3
14	Niamey (Nigelec)	32.3	24.4
15	Malanville	40.1	24.4
16	Kainji amont	-	32.2
17	Jebba Bridge	54.8	33.5
18	Baro	77.4	50.7
19	Lokoja	191.7	150.8
20	Buffle noir	1.6	1.2
21	Riao	8.3	5.5
22	Garoua	12.3	7.5
23	Wuro Boki	22.4	16.9
24	Ibi	69.3	52.8
25	Makurdi	117.2	93.1
26	Umaisha	117.9	91.6
27	Onitsha	206.8	165.1
28	Bougouni	4.3	2.2
29	Dioila	7.2	2.9
30	Douna	23.6	7.7
31	Benery Kegny	20.3	7.2
32	Kankan	6.5	4.9
33	Katsina Ala	23.3	25.1
34	Jalingo	9.9	8.0

Flow data in bcm/y

Figure 1. Niger Basin river network of hydrometric stations (Note: flow data are shown in billion m^3/year)

Arora (2002) used the aridity index, $\varphi = E_0/P$, i.e. the ratio of annual potential evapotranspiration (E_0) to precipitation (P), to assess climate change impacts on annual runoff. Simple analytic expressions based solely on the aridity index of a basin are used to estimate changes in runoff due to changes in precipitation and (temperature driven) changes in potential evapotranspiration, as a first order estimate of the effect of climate change on annual runoff. The aridity index, φ, has been shown to describe the actual evaporation ratio E/P and the runoff ratio Q/P (= 1-E/P) of catchments for a range of climatic regimes. Arora (2002) used the aridity index to obtain analytic equations, which can be used to estimate relative changes in annual runoff due to relative changes in annual precipitation and available energy, i.e. the precipitation elasticity ε_P (= [dQ/Q]/[dP/P]) and the evaporation elasticity ε_{E0} (= $[dQ/Q]/[dE_0/E_0]$) of runoff. The latter is used to derive the temperature elasticity of runoff, ε_T = [dQ/Q]/[dT/T]. Because of its mathematical convenience, we used the functional form introduced byTurc (1954) and Pike (1964), as follows:

$$E/P = [1+ \varphi^{-2}]^{-0.5} \text{ (actual evaporation ratio)}$$
$$Q/P = 1 - E/P \text{ (runoff coefficient)}$$
$$dQ/Q = (1 + \beta)\, dP/P - \beta\, dE_0/E_0$$
$$\varepsilon_P = [dQ/Q] / [dP/P) = 1 + \beta \text{ (precipitation elasticity of runoff)}$$
$$\varepsilon_{E0} = [dQ/Q] / [dE_0/E_0] = -\beta \text{ (potential evapotranspiration elasticity of runoff)}$$

$\varepsilon_T = [dQ/Q]/[dT/T] = \varepsilon_{E0}\ T/(T+17.8) = -0.60\beta$ (temp. elasticity of runoff; $T=26.6^0C$; Hargreaves, 1982, 1985)

$$\beta = [1 + \varphi^2]^{-1} / \{[1 + \varphi^{-2}]^{0.5} - 1\} = (1 + E/P)\ E/P = 2 - 3\ Q/P + (Q/P)^2$$

Thus, the (observed) runoff coefficient provides a simple initial estimator for the climate elasticities of runoff for the Niger Basin ($T=26.6^0C$), i.e.:

$$\varepsilon_P = 3 - 3\ Q/P + (Q/P)^2$$

$$\varepsilon_T = -1.2 + 1.8\ Q/P - 0.6\ (Q/P)^2$$

Results of Runoff Elasticity Analysis

We applied linear regression analysis on the historical relative variations (in %) in annual precipitation and runoff for multiple sub-catchments of the Niger Basin, and compared actual and theoretical values of the aridity index φ and precipitation elasticity ε_P based on the runoff coefficient Q/P (Fig. 2). Overall the runoff regime of the Niger Basin (i.e. for the runoff generating sub-catchment in the Upper Niger and Benue Basins) is well represented by a runoff coefficient Q/P = 0.2, which corresponds to a precipitation elasticity of runoff $\varepsilon_P = 2.44$ and temperature elasticity $\varepsilon_T = -0.86$. The former value agrees well with the results of our regression analyses, and we have thus adopted the precipitation elasticity $\varepsilon_P = +2.5$ for our study.

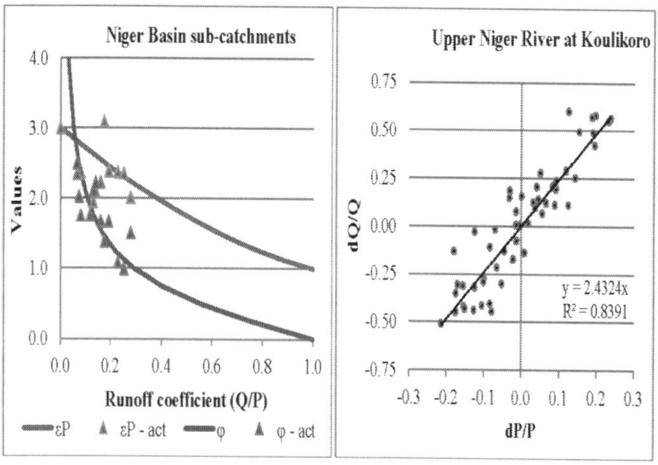

Figure 2. Theoretical and actual ε_P and φ parameter values for NRB sub-catchments, based on the runoff coefficient (left panel; blue and red lines represent theoretical values) and linear regression of historical changes in annual precipitation and runoff for the Upper Niger River at Koulikoro (right panel)

Hydrological modeling studies for similar basins in Africa generally showed lower values for the temperature elasticity of runoff, and we have thus adopted $\varepsilon_T = -0.75$[1] - . Thus, a 10% (2.7^{0}C) increase in temperature (causing a 7.5% decline in runoff) and a concomitant 3% increase in precipitation (causing a 7.5% increase in runoff) would yield no net change in runoff. Projections of future precipitation and temperature were subsequently translated into annual basin runoff for the ensemble of 38 climate projections for the 21st century, centered on 2030, 2050 and 2070 (Fig. 2), by using the precipitation - temperature - runoff regression model of the form:

$$dQ/Q_0 = \varepsilon_P \, P/P_0 + \varepsilon_T \, dT/T_0$$

The projected mean flows are essentially constant over the 21st century at an average of 2% below the 20th century mean till 2050, with an insignificant increase of 1% by 2070; the standard deviation of projected mean flows is 7% by 2030 and 13% by 2070, reflecting the increasing uncertainty in climate projections for the distant future. The probability of a decrease in average annual runoff is just over 50%. The probability of a 20% decrease in average runoff by 2050 is minimal, which could be considered as a worst case scenario for standard project economic analyses for SDAP, with 2050 as an investment horizon. The projected mean flows adhere well to the normal probability distribution (Fig. 3, right panel).

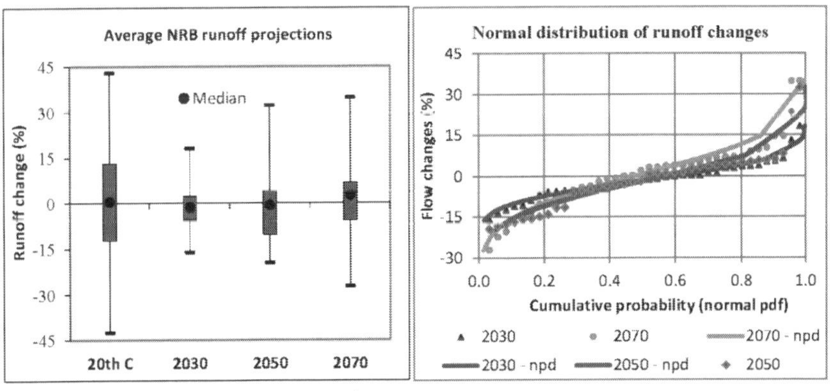

Figure 3. Runoff quartiles and probability distributions for projected runoff changes (2030, 2050 and 2070)

By 2050 the projected temperature increase (8% or 2.1^{0}C) causes an average decrease of 6% in runoff due to increased evapotranspiration, which is partially compensated by a projected 4% additional runoff due to increased rainfall, yielding a net decrease in mean runoff of only about 2%. Since annual rainfall, temperature and runoff adhere to the normal distribution, we can estimate the distribution of projected changes in runoff from the expected average changes in rainfall and temperature, and the variances of these changes (as for example available from the Climate Portal and Climate Wizard), as follows:

$E\{dQ/\mu_Q\} = \varepsilon_P \; E\{dP/\mu_P\} + \varepsilon_T \; E\{dT/\mu_T\}$; E{...} denotes the expected/projected shift in means of P and T

$CV^2(dQ/\mu_Q) = \varepsilon_P^2 \; CV^2(dP/\mu_P) + \varepsilon_T^2 \; CV^2(dT/\mu_T)$; CV(...) denotes coefficients of variation of projections

The contribution of temperature variations to the variance of runoff is negligible due to its small CV compared to the CV of changes in precipitation. This further explains that the temperature elasticity cannot be estimated through regression analysis. Probability distributions of future annual runoff can thus be based on an estimation of the average change in runoff - based on projected changes in annual precipitation and temperature and climate elasticities of runoff - and a commensurate shift of the historical distribution of annual runoff, as long as the inter-annual variability of precipitation does not change.

While all analyses pointed to a precipitation elasticity of about +2.5, the choice of the temperature elasticity or sensitivity is less certain, yet critical for the outcome of climate risk assessment. Therefore, a sensitivity analysis was done for a precipitation elasticity of +2.5 and a temperature elasticity of -1.25[2] - (instead of -0.75). Results indicate that by 2050 the increase of 8% in temperature (2.1^{0}C) would cause indeed an additional decrease of runoff by 4% due to the difference in temperature elasticities.

QUANTITATIVE CLIMATE RISKS FOR KEY WATER RELATED SECTORS

Climate risks to key sectors are estimated on the basis of projected future runoff changes, where climate risk is defined based on probabilities of selected percentage changes in performance metrics relative to baseline operations. Recall that the baseline development scenario was defined in the preceding chapter as the scenario in which the Fomi (FO), Toussa (TA) and Kandaji (KD) dams and associated irrigation infrastructure are fully implemented. As previously discussed, the relationships between relative changes in runoff and relative changes in selected performance criteria at basin level were derived from numerous Mike-Basin runs for variations in runoff between -30% and + 10% of the 20th century baseline conditions, and water demands with a 5% increase compared to the 20th century baseline (Table 3, preceding chapter). The runoff changes required to cause specific changes in selected performance indicators are shown in Fig. 11 (preceding chapter). The results shown in Fig. 11 for minimum flows are less realistic than the other results, since minimum flows are impacted mainly by changes in dry season flows and irrigation abstractions, rather than by changes in annual flows. They are indicative nonetheless of the sensitivity of minimum flows to climate changes.

Figure 4 shows probabilities of risks for the average (A) and 1/5 years (20% dry) performance of selected indicators (risk levels defined as shown in Fig. 3 of the preceding chapter), estimated as follows. The relationships between specific changes in the performance indicators and changes in the runoff required to generate those specific changes in most indicators is nearly linear in the most relevant domain of -20% to 0% of change in runoff (Fig. 11 of the preceding chapter). Thus, the performance indicators will also be distributed according to the normal probability distribution, similar to the projected changes in runoff. When we find e.g. a 25% probability that future mean flows (averaged over many years) will at least be 10% less than the present long-term mean flows, we also assume there is 25% probability that a specific performance indicator is at least xy% less than the historic indicator values; the xy% value is derived from the column '-10%R' in the performance matrix of Table 3(preceding chapter). In reverse, to estimate the probability that e.g. projected hydro-energy generation will at least be 20% less than at present, we determine the runoff reduction required to cause 20% less hydro-energy (16% per Fig. 11 of the preceding chapter), and estimate the probability of such runoff reduction based on the normal distribution for the projected future runoff (2% in 2030 and 10% in 2070).

Figure 4 shows specific percentiles (5, 25, 50, 75 and 95%) of the selected performance indicators. The concerned percentiles of runoff changes were first determined from the normal probability distribution of projected future runoff changes; then the corresponding percentiles of each performance indicator were interpolated from the performance matrix in Table 3 of the preceding chapter.

The probability that by 2050 or 2070 one may see decreases of more than 20% (significant risk) in the various performance criteria is only 5 to 15% or less, other than for the minimum environmental flow conditions in the Inner Delta and Middle Niger, which are under severe risk. The probability that one may not be able to maintain the required minimum flows under the fully developed FO-TA-KD scenario is 100%. This will be primarily caused by (projected) increased water abstractions for irrigation at the large Office du Niger irrigation scheme in Mali. Hydro-energy, navigation and flooding of the Inner Delta vary similar to runoff variations, with runoff elasticities of about 1 to 1.2. There is 25 - 35% probability that by 2050 these sectors could suffer performance decreases above 10% (moderate risk). The probability that the performance decreases in these sectors would be between 10 and 20% is about 20% and the probability of performance decreases exceeding 20% is between 5 and 15%. Impacts on irrigated agriculture are minimal under the prevailing priority of water allocation to agriculture.Figures 3 and 4 serve to illustrate the main findings and conclusions of this CRA, as highlighted in the following.

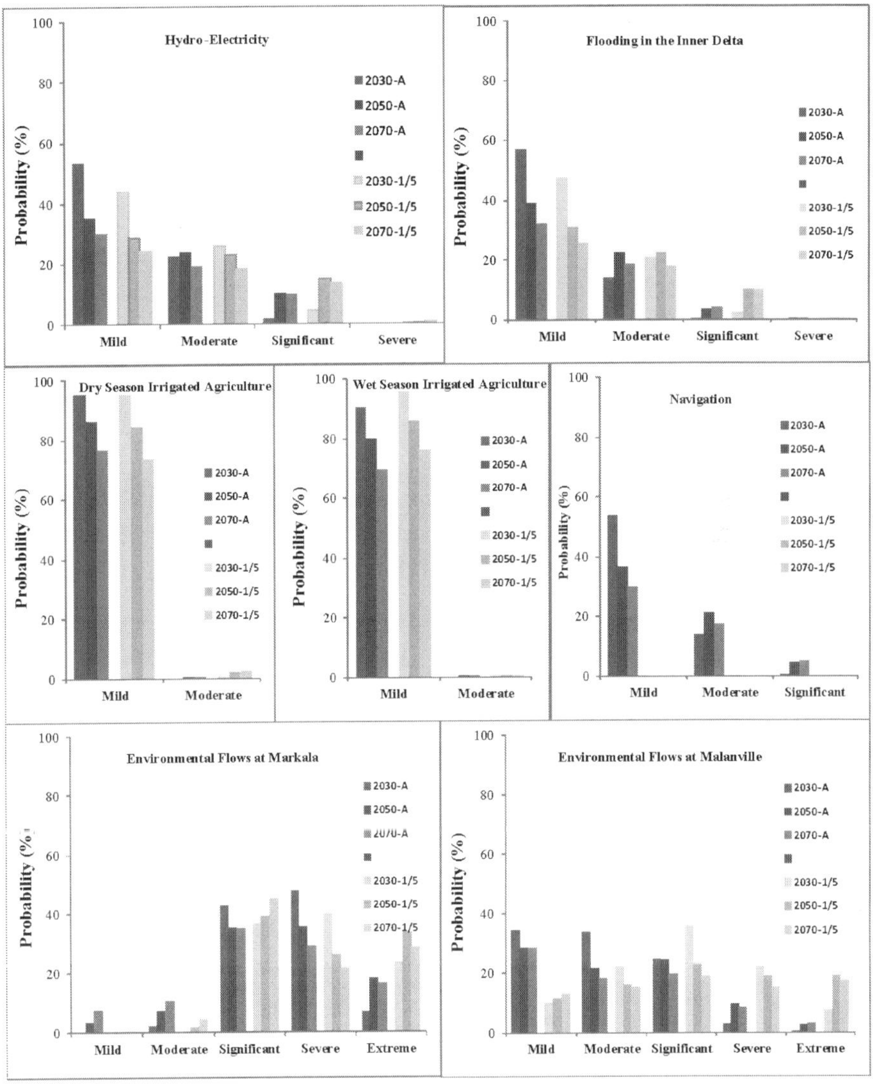

Figure 4. Probabilities of risks for average (A) and 1/5 years (20% dry) performance of selected indicators; risk levels are defined in Fig. 3 of the preceding chapter.

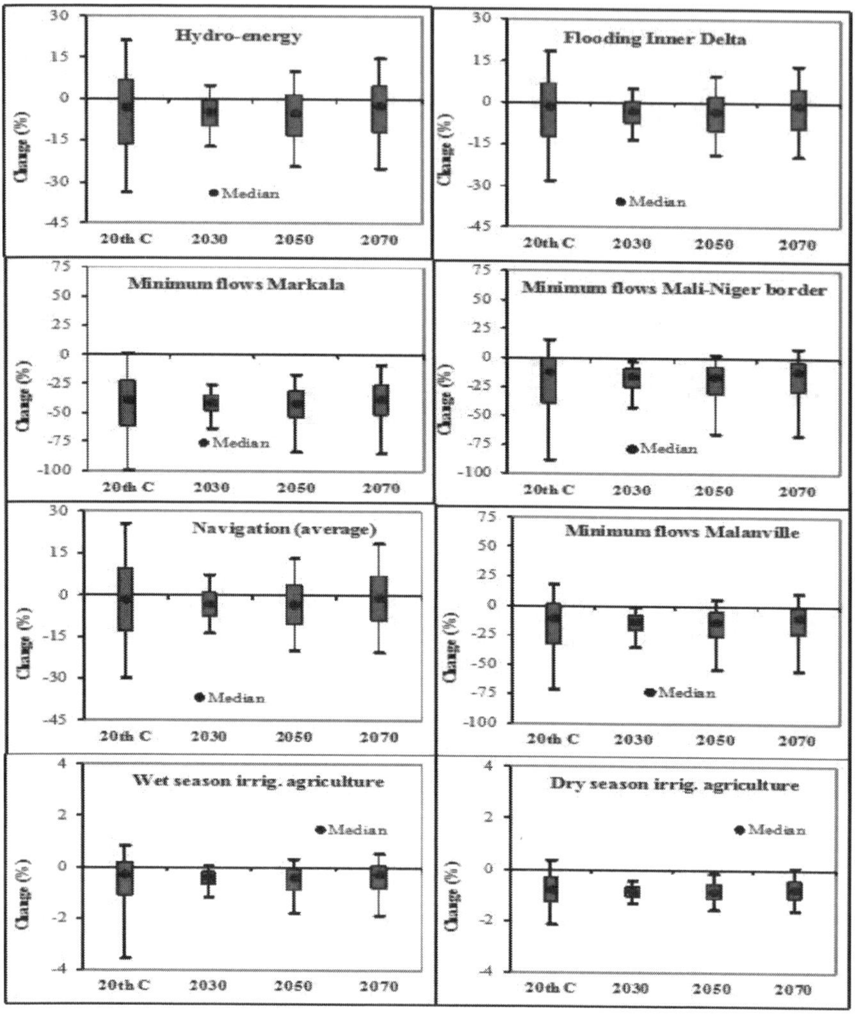

Figure 5. Percentiles (5, 25, 50, 75 and 95%) of selected performance indicators for an average year (Note: Minimum flow data refer to 10-day average flows)

Sensitivity of Selected Sectors to Climate Change

Irrigated agriculture is insensitive to projected climate changes; SDAP and particularly the construction of Fomi dam is an effective adaptation measure. Under SDAP total agricultural output in the NRB is projected to increase by nearly 450%. The current water allocation rules prioritize irrigation water demands in order to secure food production and alleviate poverty in the long term. This makes irrigated agriculture in the NRB insensitive to decreased runoff as a result of the projected climate changes, but it places pressures on the existing reservoir systems to provide a reliable water supply for hydro-electricity production, navigation, and environmental flows. These sectors are found to be more vulnerable to the risk of decreased runoff. Under the current water allocation rules, mild agricultural production decreases are likely to occur, but would generally be less than a few percentage points. As long as reservoirs can be fully replenished during the rainy season, climate change has little impact on the supply side of irrigation during the dry season. Indeed, by regulating the variability of the Upper Niger flow, SDAP and particularly the construction of Fomi dam will be good insurance for the protection of irrigated agriculture against the potential negative impacts of climate change.

Climate change impacts on hydro-energy, navigation, and Inner Delta Flooding are projected to be mild to moderate (less than 20% decrease). Hydro-energy, navigation and flooding of the Inner Delta vary similar to runoff variations, with runoff elasticities of about 1 to 1.2[2]. There is 25 to 35% probability that by 2050 these sectors could suffer performance decreases more than 10%, and the probability of performance decreases exceeding 20% is only between 5 and 15%. Hence, risks of climate change impacts on these sectors will be mild to moderate. These impacts are similar to the declines in navigation and Inner Delta flooding caused by the full implementation of the FO-TA-KD development scenario (10-20%).

Climate change impacts on minimum flows can be severe. Minimum flows entering the Inner Delta (downstream of the Office du Niger irrigation scheme in Mali) are sensitive to an increase in water demands by for example 5% at the Office du Niger (ON) irrigation scheme, due to increased future temperatures and evapotranspiration. The design of SDAP has planned irrigation development and abstractions during dry season irrigation at the ON such that minimum flows at Markala dam can just be maintained at the present ON water demands per hectare for dry season irrigation. The probability that by 2050 - under the FO-TA-KD development scenario - one may not be able to maintain the required minimum flows through the Inner Delta and the Middle Niger to Malanville at the Niger-Nigeria border is 100%.

Minimum environmental flows during the dry season are the most sensitive to climate change(particularly with increased irrigation water demands) under the present water allocation rules, with runoff elasticities on the order of 2.5 to 4.0[3] - . By 2050 there is a 50% probability that once in 5 years the 10-day average minimum inflow to the Inner Delta will be less than 25 m^3/s; i.e. less than 60% of the adopted norm of 40 m^3/s.

Relative decreases of minimum flows at the Mali-Niger border, Niamey and Malanville are slightly less severe.

Adaptations required for enhancing minimum Niger River flows. Minimum flows are severely impacted by increased water demands of the Office du Niger (ON) irrigation scheme due to increased evapotranspiration. Should higher priority be accorded to the sustenance of minimum flows, as per the Water Charter agreed between the NBA member countries - dry season irrigated agriculture could become moderately sensitive to climate change impacts, since the abstractions for irrigation at ON – as planned under the FO-TA-KD scenario - would need to be reduced in favor of releasing more water into the Inner Delta. This potential problem can be addressed by measures to increase the present low dry season irrigation efficiency at ON (about 30%), and/or by slightly reducing the future dry season irrigated areas, such that minimum flows can also be maintained in the future. Future abstractions during the critical low flow period February to May would by 2050 amount on average to 425 m^3/s, and would need to be reduced by about 25 m^3/s (6%) to avoid unacceptable minimum environmental flows passing Markala dam. This can be achieved by implementing irrigation system rehabilitations and water management measures aimed at increasing the dry season irrigation efficiency at ON by about 2%; presently the dry season irrigation efficiency at ON is less than 30%. Other options to reduce abstractions by at least 6% during the dry season would be to reduce the acreage projected to be cropped in the dry season by 6%, or to reduce dry season rice planting in favor of an extension of less water demanding horticulture and other non-rice crops.

Worst case scenario for project economic analysis. The probability of a 20% decrease in average runoff in the Niger Basin is minimal, which could thus be considered a worst case scenario for the standard economic analysis of infrastructure projects under SDAP. For example, a detailed economic analysis of the Kandadji hydro-power and irrigation project - located on the Niger River in Niger near the Malian border - was performed as part of project appraisal (World Bank, 2012), yielding an Economic Internal Rate of Return (EIRR) of 13.5% for the base case (2005 hydrology). Sensitivity analyses showed that the EIRR would reduce to 12.4% for an overall decline of 20% in the runoff of the upstream basin, still above the World Bank's financing threshold rate of 12%. The runoff elasticity of EIRR was assessed at about 0.4, i.e. a 20% decrease in runoff causes an 8% decrease in the EIRR (1.1 percentage points down from 13.5%). Thus a 30% decline in average runoff would push the EIRR below the threshold rate of 12%.

Economic Analysis of Sdap under Climate Change Conditions

Potential climate change impacts on the economic performance of SDAP are modest. The economic analysis of climate change impacts on the full SDAP investment plan has focused on hydropower development with associated irrigation development as the main drivers for SDAP. Various proposed Run-of-River (R-o-R) hydropower schemes were also included. Primary economic benefits from the FO-TA-KD Program and R-o-R schemes will accrue from the combination of hydropower generation and new schemes brought under irrigation. Secondary benefits include provision of water supply (particularly for towns and villages along the Middle Niger River), environmental benefits of wetlands (reservoir surface areas and newly irrigated rice planted areas), fish production in the new reservoirs, ecosystem regeneration in the Niger Valley, and livestock production. Economic and environmental losses will occur in the Inner Delta (rice production, environmental benefits of wetlands, fisheries and livestock), related to navigation in the Middle and Lower Niger, and due to reduced hydro-energy generation at Kainji and Jebba dams in Nigeria caused by upstream water diversions for irrigation and evaporation losses from the new reservoirs. Losses due to climate changes, which would also occur under the prevailing (2005) "natural conditions", such as losses of Inner delta flooded areas and losses of hydro-energy at Kainji and Jebba dams due to climate change, have been excluded from the analysis of climate change impacts on the SDAP Program. Results of the economic analysis for the SDAP, in terms of EIRR, are shown in Table 1, for the baseline (2005) hydrological conditions and for the situation with a 20% flow reduction due to climate change during the entire life of the SDAP project components (2015 – 2050).

Table 1. Estimated EIRR for various SDAP development scenarios, with/without impacts of climate change

Development scenario	EIRR (2005 hydrology)	EIRR under 20% runoff reduction	Runoff elasticity of EIRR
FO-TA-KD + R-o-R	13	11.3	0.64
FO-KD + R-o-R	14.6	13	0.55
FO-TA-KD; no R-o-R	9.5	7.9	0.84
FO-KD; no R-o-R	11.7	10.2	0.64
FO-TA-KD + R-o-R; no irrigation development	14.1	12.3	0.64
FO-TA-KD; no R-o-R; no irrigation development	10.1	8	1
KD + 45,000 ha irrigation (no R-o-R)	13.5	12.4	0.42

Potential climate change impacts on the economic performance of the SDAP Program are assessed as mild (<10%) to moderate (<20%). Under the above 'worst case' scenario of 20% reduction in the long-term average basin runoff, the projected climate change impacts on irrigated agriculture, navigation, minimum flows and flooding of the Inner Delta have only a minor impact on the overall economic performance of the SDAP. In economic terms the only significant impact of reduced runoff due to climate changes would stem from its direct impact on hydro-energy production. In the 'worst case scenario', with the 20% flow reduction imposed as from 2015, one can expect a 1.7% reduction in the EIRR of 13% for the SDAP under the base case (2005) hydrological conditions. The runoff elasticity of the EIRR for the SDAP is about 0.6, compared to 0.4 for the Kandadji (stand-alone) Program. The investment in Run-of-River hydropower schemes appears to be the most beneficial component of the SDAP, since it increases the EIRR of the FO-TA-KD scheme by 3.5%. It allows the SDAP to maintain a robust EIRR above 12%, and even more than 11% in the 'worst case' scenario. Without R-o-R schemes the EIRR drops well below 10% under conditions of 20% reduction in average runoff due to climate change.

Irrigated agriculture achieves an adequate EIRR as long as the cost of the new dams are treated as sunk cost, i.e. are fully charged to hydropower development. Losses to dry season irrigated agriculture under climate change, estimated at 8% for an average runoff reduction of 20%, cause only a loss of 0.1% of the EIRR under the 'worst case' runoff scenario. Similarly, the impacts of losses due to reduced flooding of the Inner Delta (including losses to wetlands, fisheries, floating rice production and livestock) caused by Fomi dam and the planned large extensions of the Office du Niger irrigation schemes have no significant impact on the EIRR, causing only a loss of 0.2% in the EIRR. Impacts of climate change on these losses are similar under the present (2005) as well as future (FO-TA-KD) hydrological conditions, and were thus excluded from the SDAP economic analysis. Finally, losses to navigation under the FO-TA-KD scenario cause only a reduction of 0.2% of the EIRR.

CONCLUSIONS

While the risk of climate change has generated many policy and adaptation initiatives aimed at combating it, few of these initiatives are based on quantitative assessments of expected impacts to specific sectors. We argue that such assessments are necessary to convince decision makers and as a means of prioritizing and targeting scarce resources at the most vulnerable sectors or those sectors likely to create the largest domino effect. This study represents an important contribution towards that goal. Beginning with a bottom-up definition of risk, we estimated the expected magnitude of climate change, principally precipitation and temperature changes, translated those changes into runoff changes, and then estimated the impacts of projected runoff changes on key development sectors in the Niger Basin with respect to thresholds of changes in specific performance indicators.

The results show that climate change impacts in the Niger Basin will be generally mild (<10%) to moderate (<20%). By 2050, average annual temperatures in the Niger Basin would rise by 2.1^{0}C, an 8% increase over baseline values. The increase in temperature will increase evapotranspiration, leading to an average decrease of 6% in runoff. However, a projected 4% additional runoff due to increased rainfall will partially offset the decreased runoff, producing a net decrease in mean runoff of only about 2%. It is notable that the magnitude of these changes pale in comparison to the historical patterns of variability in precipitation and runoff, observed over the NRB in the 20th century. Even so, the results should be interpreted cautiously. In general, while climate projections do a good job on simulating the mean, they are generally incapable of reproducing variability around that mean. Thus, whereas the change in the mean may be small, variability around the mean may trigger shocks to economic and agro-ecological systems disproportionate to the change in mean conditions.

With specific respect to the Niger Basin SDAP and under the current water allocation rules, irrigated rainy season agriculture ('hivernage') remains insensitive to climate change even under severe reductions of runoff (e.g. up to 30%). Dry season irrigated agriculture ('contre-saison') and the sustenance of environmental flows during the low flow season are interlinked and respectively mildly and extremely vulnerable to severe reductions of runoff. However, with (i) the implementation of Fomi, Taoussa and Kandadji dams, along with (ii) optimal basin-wide reservoir management, (iii) increased irrigation efficiencies, particularly at the Office du Niger, (iv) gradual shifts from dry season rice to less water demanding non-rice dry season crops, and (v) other similar adaptation measures to improve water use efficiencies in particularly dry season irrigated agriculture, these 'sectors' can be well protected from the impacts of significant climate change.

Severe reductions in runoff would cause equally severe reductions in generated hydro-energy, navigation and flooding of the Inner Delta (projected to be mild to moderate under the present climate change projections). Such severe future impacts on (particularly) hydro-energy generation can only be minimized by reducing rainy season irrigated agriculture in the basin and/or by the construction of additional storage reservoirs along with hydro-energy generation facilities in the Upper Niger Basin and in Nigeria, particularly in the Benue Basin, which has mostly untapped hydro-power potential.

It is important to emphasize that in this report the risks are calculated based on the long term (i.e. 30 years) shift in mean precipitation and temperature values; they do not account for inter-annual or decadal variability. We do not consider this a serious limitation since the baseline period (1966-1988) - used also for the design of the FO-TA-KD scenario and simulated with the Mike Basin model - comprises arguably the most variable period in the historical data. It is noteworthy that the NRB has historically experienced runoff shortages greater than the projected levels. Thus, the historical experience provides an analogue for dealing with future climate; for water managers and farmers who do not know what to expect in the upcoming rainy season, managing the impacts of intra-seasonal, inter-annual variability of climate would be the priority to start with. Managing the near-term climate variability has also the potential to better prepare for dealing with long term-climate change impacts. Therefore, the use of seasonal to inter-annual hydrologic forecasts in reservoir operations and planning could be an important adaptation opportunity.

Notes

[1] - The temperature elasticity of -0.75 corresponds to a temperature sensitivity of run equal to -0.75/T, or approximately -3% runoff per 0C temperature increase.

[2] - The temperature elasticity of -1.25 corresponds to a temperature sensitivity of runoff equal to -1.25/T, or approximately 5% decrease in runoff per 0C increase in temperature, which is well above the values mostly found in literature.

[3] - The large runoff elasticities of minimum flows indicate that moderate decreases in annual runoff can cause severe reductions in the minimum flows at Markala and further downstream on the Middle Niger.

ACKNOWLEDGEMENTS

This study was done as part of the Climate Risk Assessment (CRA) for the Niger Basin, a joint initiative of the Niger Basin Authority (NBA) and the World Bank to assess the risks from climate change to the performance of NBA's Sustainable Development Action Plan (SDAP) for the Niger Basin.The aim of this initiative is to build resilience to climate risks into the SDAP. The Niger CRA study is supported by grants from (i) the Bank-Netherlands Partnership Program (BNPP) Trust Fund, (ii) the Trust Fund for Environmentally and Socially Sustainable Development (TFESSD) funded by Finland and Norway, (iii) the Norwegian Trust Fund (NTF) and (iv) the Trust Fund for Integrated Land and Water Management for Adaptation to Climate Variability and Change (ILWAC) funded by Denmark. We gratefully acknowledge the NBA Observatory for making hydro-meteorological and runoff data available, as well as providing valuable comments and directions for the study. We extend our thanks to Dr. Amal Talbi for managing this CRA study for the World Bank, and to the World Bank's Water Partnership Program (WPP) and Water Unit for organizing the special session S3 of HydroPredict2012 (Vienna, September 2012) on 'Choosing Models for Resilient Water Resources Management' (Grijsen et. al, 2013), and for giving permission to present the results of this CRA work in this book.

REFERENCES

1. Arora, V.K., 2002: The use of the aridity index to assess climate change effect on annual runoff, Journal of Hydrology, 265, p. 164–177.
2. Deksyos T. and T. Abebe, July 2006: Assessing the impact of climate change on the water resources of the Lake Tana sub-basin using the Watbal model. CEEPA Discussion Paper No. 30, Special Series on Climate Change and Agriculture in Africa. Centre for Environmental Economics and Policy in Africa, University of Pretoria, South Africa.
3. Gleick, E. H., 1986: Methods for Evaluating the Regional Hydrological Impact of Global Climatic Changes, J. of Hydrology 88, 97-116.
4. Gleick, E. H., 1987: Regional Hydrologic Consequences of Increases of Atmospheric CO2 and Other Trace Gases, Climatic Change 110, 137-161.
5. Grijsen, J.G. and C. Brown, 2013: Climate elasticity of runoff and climate change impacts in the Niger River Basin; Niger River Basin Climate Risk Assessment, a joint initiative of the Niger Basin Authority (NBA) and the World Bank (not yet disclosed).

6. Grijsen, J.G., C. Brown, A. Tarhule: Climate Risk Assessment for Water Resources Development in the Niger River Basin; HydroPredict'2012 - International Conference on Predictions for Hydrology, Ecology and WRM, Special Session S3 – Choosing models for resilient water resources management, Vienna; World Bank WET/WPP/TWIWA publication (in print).
7. Hirabayashi, Yukiko, Shinjiro Kanae, Ken Motoya, Kooiti Masuda and Petra Doll, 2008: A 59-year (1948-2006) global near-surface meteorological data set for land surface models. Part I: Development of daily forcing and assessment of precipitation intensity. Hydrological Research letters 2, 36-40.
8. Karl, T.R. and W.E. Riebsame, 1989: The impact of decadal fluctuations in mean precipitation and temperature on runoff: A sensitivity study over the United States, Climatic Change 15: 423-447.
9. Legates, D., H. Lins, H., and G. McCabe, 2005: Comments on "Evidence for global runoff increase related to climate warming" by Labat et al., Advances in Water Resources, 28, 1310–1315, 2005.
10. Pike, J. G., 1964: The estimation of annual runoff from meteorological data in a tropical climate. Journal of Hydrology, Amsterdam, 2, 116–123.
11. Risbey, J. S., and Entekhabi, D, 1996: Observed Sacramento basin streamflow response to precipitation and temperature changes and its relevance to climate impact studies, Journal of Hydrology.
12. SNC Lavalin International, February 2007: Strategic/Sectoral, Social and Environmental Assessment of Power Development Options in the Nile Equatorial Lakes Region, Final Report, Chapter 12 and Appendix K.
13. Strzepek K. M. and A. McCluskey, July 2006: District level hydroclimatic time series and scenario analyses to assess the impacts of climate change on regional water resources and agriculture in Africa; CEEPA Discussion Paper No. 13, Special Series on Climate Change and Agriculture in Africa.
14. Turc, L., 1954: Le Bilan d'eau des sols. Relation entre les précipitation, l'évaporation et l'écoulement. Annales Agronomique, 5, 491–595.
15. Vogel, R.M., I. Wilson and C. Daly, May/June 1999: Regional regression models of annual stream flow for the United States, Journal of Irrigation and Drainage Engineering.
16. Wigley, T. M. L. and E.D. Jones, 1985: Influences of Precipitation Changes and Direct CO2 Effects on Streamflow, Nature 314, 140-152.
17. World Bank, 2012: Project Appraisal Document on the 1st part of the 2nd phase of the Niger Basin Water Resources Development and Sustainable Ecosystems Management Project (WRD-SEM APL 2A).

CITATION

G. Grijsen, A. Tarhule, C. Brown, Y.B. Ghile, Ü. Taner, A. Talbi-, Jordan, H. N. Doffou, A. Guero, R. Y. Dessouassi, S. Kone, B., Coulibaly and N. Harshadeep (2013). Climate Risk Assessment for Water Resources Development in the Niger River Basin Part II: Runoff Elasticity and Probabilistic Analysis, Climate Variability - Regional and Thematic Patterns, Dr. Aondover Tarhule (Ed.), ISBN: 978-953-51-1187-0, InTech, DOI: 10.5772/56708. Available from: http://www.intechopen.com/books/climate-variability-regional-and-thematic-patterns/climate-risk-assessment-for-water-resources-development-in-the-niger-river-basin-part-ii-runoff-elas

CHAPTER 9

Development and Uptake of Scenarios to Support Water Resources Planning, Development and Management – Examples From South Africa

Nikki Funke[1], Marius Claassen[1] and Shanna Nienaber[1]

[1]Natural Resources and Environment Unit, Council for Scientific and Industrial Research, Pretoria, South Africa

INTRODUCTION

The international agenda on water resources development reflects societal needs, political agendas, economic realities and the state of resources. The industrial revolution, which started in the 18thcentury, brought social and economic prosperity but also marked a major shift in humanity's impact on the earth's systems. This shift is now referred to as the Anthropocene [1], where humans have brought such vast and unprecedented changes to the planet that this era represents a new geological time interval [2]. Societal needs have shifted since the 1940s from a need for modest food production to a need for increased agricultural productivity that has been met by high yield crops, the use of pesticides, the application of fertiliser and advanced agricultural techniques. This development has averted food shortages, but has also resulted in humanity having to pay a heavy price in terms of increased water use and energy consumption, as well as environmental degradation [3].

From the early 1970s a series of events and key documents has promoted an integrated approach to sustainable development. The 1972 United Nations Conference on the Human Environment considered the need for a common outlook towards the preservation and enhancement of the human environment [4]. The World Commission on Environment and Development advanced this agenda in their report 'Our Common Future', with an emphasis on sustainable development promoting harmony among human beings and between humanity and nature [5]. The International Conference on Water and the Environment that took place in Dublin in 1992 resulted in the development of four guiding principles [6]. These principles, commonly referred to as the Dublin principles, state that: water is a finite resource with economic value and social implications; local communities must participate in water management; water resources management must be developed within a set of policies; and the role of rural populations and women should be recognised. This led to the Rio Declaration and the adoption of Agenda 21, which is a comprehensive plan of action to be implemented globally, nationally and locally in every area in which humanity impacts on the environment [7]. This declaration subsequently became the blueprint for sustainable development world-wide [8].

Uncertainties about societal, economic, political and environmental aspects have proved to be a considerable obstacle to the implementation of sustainable development. Here follow a few examples of such uncertainties. In 1980, the World Development Report of the United Nations [9] estimated that the world population would reach 6.029bn by the year 2000. Five years later, the estimate was updated to 6.088bn [10], with further updates at five yearly increments resulting in estimates of 6.194bn and 6.123bn [11, 12]. The actual population in the year 2000 turned out to be 6.188bn [13]. Future economic development is also uncertain, with the annual growth in Figure 1 showing how the world average varies significantly between years and also how the growth of individual countries (South Africa in this case), does not necessarily follow the global trend and is even more variable between years.

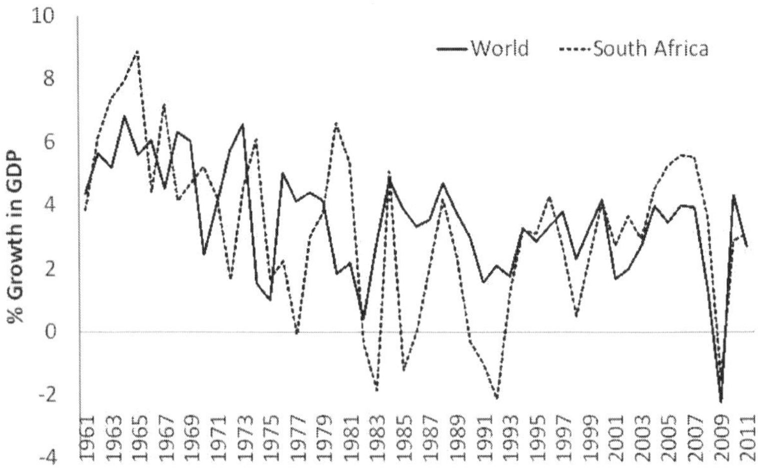

Figure 1. Annual economic growth between 1961 and 2011 [13].

Environmental conditions also vary significantly over time and space, with Figure 2 illustrating the annual deviation of rainfall over southern Africa. This uncertainty is exacerbated by climate projections, which suggest that freshwater resources are vulnerable and have the potential to be strongly impacted by climate change, with wide-ranging consequences for human societies and ecosystems [14].

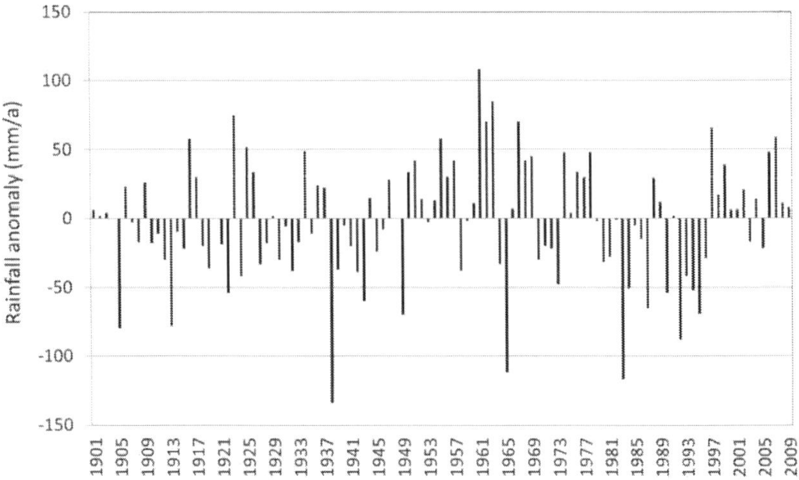

Figure 2. Annual rainfall anomalies for the southern African region (1901-2009; Adapted from [15]).

A question that emanates from the realisation that we live in a changing world where change is unpredictable is, 'How do we plan for the future?'

Water use in South Africa was first regulated through the Irrigation and Conservation of Waters Act (Act No. 8 of 1912), which managed the use of water from public streams for domestic, irrigation and industrial purposes [16]. The Water Act (Act No. 54 of 1956) further regulated water use by providing for the control of water pollution and the more effective protection of water resources. The variable distribution of water required the development of infrastructure to capture, store and distribute water. The subsequent expansion of mines, industries and urban areas created a demand for further infrastructure development. When this demand further increased and the social and economic issues in South Africa became increasingly complex in the 1990s as the country was transitioning from apartheidto democracy, a shift in thinking was required. As it became clear that engineering solutions to increase water supply were not sustainable, a holistic strategy to meet future needs became more popular [17].

The new National Water Act (Act No. 36 of 1998) [18] emphasised water resources management at national and catchment scales, made specific provisions for the protection of water resources, established mechanisms to ensure equitable and efficient water use and promoted participatory management. The National Water Resources Strategy [19] addressed the balance of future water supply and demand by establishing scenarios. The demand scenarios were based on population growth by 2025, with the high population scenario at 54 million people and the low population scenario at 50 million people. It also established economic growth scenarios, with the upper scenario assuming 4% growth in GDP and the less favourable scenario assuming 1.5% [19].

While there has been much progress in water infrastructure development for services (public benefit), the backlog in issuing water use licenses (mostly for private benefit) stood at 4 318 in 2011 [20]. The protection of water resources has suffered as a result of the government's drive to achieve social and economic development, with South Africa ranked 128 out of 132 countries in the Environmental Performance Index [21]. The National Water Act provides for a balance of responsibilities, ranging from the Minister and Director General at the national level, to Catchment Management Agencies (CMAs) at the basin level and Water User Associations (WUAs) at a sub-basin level. Progress has been slow as after 14 years after the promulgation of the Water Act, only two CMAs (out of the 19 intended) have been established [20]. It can be argued that many hurdles have to be overcome to fully realise cooperative governance for

Integrated Water Resources Management (IWRM), with inadequate human and institutional capacity being one of the main factors limiting the efficient management of water resources in South Africa [22]. To illustrate this point: the country's Department of Water Affairs (DWA) reported having 4 286 people in its employment in September 2010, while 1 155 posts were vacant at the time [20].

From the discussion above it becomes clear that we live in a world with social, economic and environmental conditions that are variable and difficult to predict, and the water sector is no exception. This uncertainty provides a challenging environment for policy and institutional development. Scenarios are one way of attempting to achieve a desired outcome in an uncertain and variable future [23]. The rest of this chapter will examine the research question, 'How are scenarios able to achieve impact in an uncertain world, with a particular focus on water resources planning, development and management?' The body of this chapter focuses on the research method, presents an overview of scenario development and the importance of scenario development and how they facilitate more effective water resources planning, development and management, focuses on a few select South African scenarios and the impact they have had and then turns to discussing the impact of scenarios in general. The conclusion wraps up the learning from this chapter and suggests a way forward in terms of future research and designing scenarios for impact.

THE ABILITY OF SCENARIOS TO ACHIEVE IMPACT IN AN UNCERTAIN WORLD WITH A FOCUS ON WATER PLANNING, DEVELOPMENT AND MANAGEMENT

The authors of this chapter conducted an exploratory study on the ability of scenarios to achieve impact in an uncertain world, with particular reference to water planning, development and management. They conducted a review of scenario planning literature in the water and other sectors, and also considered literature focusing specifically on the impact of scenarios. The authors also considered literature on the impact of scientific research and on the science-policy interface. This was accompanied by a search of major databases (e.g. Google Scholar, EBSCO Host and Scopus) to determine where and how the four scenarios discussed in this chapter have been cited. In addition, the authors interviewed selected stakeholders in the water and other sectors who are likely to have been exposed to scenarios and who may use scenarios when making decisions in their workplace.

Scenarios and Their Importance in the Water Sector

The concept of scenario planning has its origin in military applications, with the US Air Force developing 'scenarios' of what the enemy might do and preparing alternative strategies. It was thus aimed at achieving a desired outcome in an uncertain future [24]. At the end of the 1940s, researchers at the RAND Corporation started to investigate the scientific use of expert opinion in planning for the future [25]. The Royal Dutch Shell company employed scenario tools to good effect in the 1970s, when they improved their size and profitability by being prepared to act quickly during the oil price shock of 1973 [26]. In summarising definitions of scenarios, scenarios can be described as a narrative description of a possible state of affairs or development over time, that they are useful to communicate speculations about the future to promote discussion and feedback, and that they can dramatise trends and alternatives, explore the impacts and implications of decisions, choices and policies, and provide cause-and-effect explanations [24].

Clem Sunter is credited with popularising the use of scenarios in South Africa, with 'The World and South Africa in the 1990s', which describe the 'High Road' and 'Low Road' scenarios [27]. The publication was based on work from Anglo American Corporation teams in London and Johannesburg. Subsequently, Adam Kahane facilitated a process that became known as the Mont Fleur scenario project, which was launched in 1992. It explored the question of 'What will South Africa be like in the year 2002?' These scenarios were arrived at collaboratively by a very broad group [28]. The Department of Arts, Culture, Science and Technology (DACST) also deployed scenarios and technology foresighting in the development of South Africa's National Research and Development Strategy, with Kahn initiating and leading the development of the South African National Research and Technology Foresight Project [29]. The Dinokeng team [30] developed '3 Futures for South Africa', which characterised future scenarios based on the effectiveness of the state and the engagement of society. Some of the recent scenario projects in the water sector include the World Business Council for Sustainable Development report on 'Business in the World of Water: WBCSD Water Scenarios to 2025' [31], and the Global Research Alliance (GRA) report on 'Science and Technology-based Water Scenarios for sub-Saharan Africa' [32].

The Importance of the Use of Scenarios in Water Resources Planning, Development and Management

Scenarios are important and useful to water resources planning, development and management in a number of ways. In the South Africa context, in particular, scenario development processes have been instrumental in initiating strategic conversations among scenario workshop In the South African context, (e.g. the transition from apartheid to democracy), and have helped develop a common language among people with widely divergent views [28]. Those involved in scenario development processes may be inspired to think more broadly about the future and the forces creating it. They may also realise how their particular actions may help to create a desired future [33]; and they may have suggestions about which options exist to direct target audiences on to a desirable path [28]. The knowledge that scenarios generate can therefore potentially empower role players in the water sector and other sectors to engage in participative governance by equipping them with insights into potential futures they may face, and making them aware of the implications of certain decisions, behaviours and actions [23]. Finally, the advantage of communicating scenarios as stories is that they have the psychological impact that other more academic means of communication, for example, graphs and equations, lack. Stories give order and meaning to events, which is crucial for imagining future possibilities [34].

Some South African Scenarios: Overview and Impact

The discussion in this chapter and the research question were inspired by the development of the Water Sector Institutional Landscape by 2025 scenarios. These scenarios were the main output of a research project led by the authors. In particular, the authors are interested in how these scenarios could be used by potential end-users. Given this question and the importance and potential usefulness of scenarios in facilitating decision-making in a context of uncertainty, it becomes important to reflect on some examples of scenarios that have been developed in South Africa at different points in history and to learn from the impact they have had on different sectors, including the water sector. These scenarios are discussed in chronological order. The section starts with the High Road/Low Road scenarios that were developed late in the apartheid era and on the cusp of South Africa's transition to democracy. Secondly, the Mont Fleur scenarios, which were developed during the democratic negotiations, are discussed. Thirdly, the section focuses on the Dinokeng scenarios that were developed in 2009, the year a new president came to power and a serious economic crisis shook the world. The section concludes with the

Water Sector Institutional Landscape scenarios that focus on potential futures of the South African water sector in 2025.

Overview and Process

The High Road/Low Road scenarios were an initiative by the Anglo American Corporation in the early 1980s and aimed to look into some less conventional approaches to business planning and future investment decisions, given the international economic turbulence of the 1970s and the resultant slump in commodity markets. During this time, South Africa's economic performance was poor and several events resulted in the country becoming increasingly isolated and the government resorting to a rule of force. Careful and gradual reforms by the apartheid government in the middle to late 1980s and increasing attempts by members of the white establishment to reach out to black leaders in exile, led to the eventual unbanning of the African National Congress (ANC) and the release of Nelson Mandela in 1990 [35].

The scenarios involved a large-scale exercise with numerous contributors, notably Pierre Wack and Ted Newland, as well as Clem Sunter. Most of the effort went into developing global scenarios which were based on the analysis of key 'drivers' (for example, demography, technology and societal values) of developments in Japan, the USA and USSR (then regarded as the main players of the world economy), and also the ingredients for success of 'winning' nations and world class companies. This work then provided the basis for the South African scenarios. In essence, these scenarios focus on the choice the country was facing to either (through consultation and negotiation) travel on the 'High Road' to a non-racial democracy and increasing prosperity, or, to continue on the 'Low Road' of confrontation, conflict and falling incomes (as a repressive, centralised society and controlled economy) and ending up as a 'waste land' [35].

The scenarios conclude with the need for a 'common vision' to help launch South Africa into the more desirable 'High Road' scenario. This common vision entails putting South Africans first (looking beyond different races and groups), to turn the country into a 'winning' nation and to work towards achieving a certain income per head, all of which would be reached through negotiation [28].

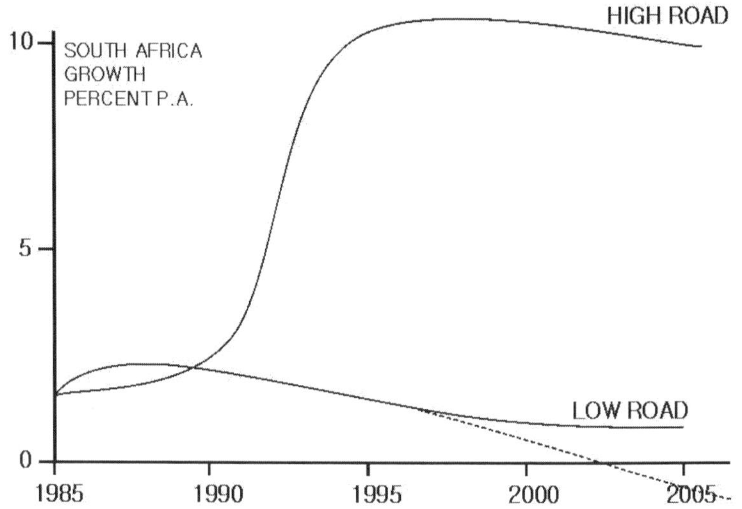

Figure 3. The High Road/Low Road scenarios depicting two possible future trajectories for South Africa [27]

Dissemination and Impact

Within a year, starting in 1986, Clem Sunter presented the 'High Road/Low Road' scenarios to 230 (mostly white) audiences at various levels of society, thereby reaching between 25 000 to 30 000 people [35]. Senior politicians of the ANC were also one of Sunter's audiences before the eventual negotiated settlement was reached [28]. The message of the scenarios seems to have made a big impression on the audiences as it was ultimately positive and encouraged people in the country to take it into their own hands to get on to the 'High Road', without being prescriptive about how this should be done [35]. In particular, the High Road/Low Road scenarios also seem to have contributed somewhat to the shift in thinking in government circles, and indeed as supporting evidence for a need for change, which eventually brought about a political transition. In conclusion then, the High Road/Low Road scenarios started out as a corporate scenario project and resulted in a brilliant communication exercise, both in terms of content and style of presentation, that reached thousands beyond the initial intended audience and paved the way for more prominent South Africa scenario exercises to come [35].

In terms of uptake in the scientific and decision-making community, Clem Sunter's book 'South Africa and the World in the 1990s' has been widely cited and includes discussions of a range of topics. These include reflections on various elements of the political and economic transformation of South Africa, the future of Africa, scenario development and planning and globalisation. The citations include a variety of different sources, including books, journal articles, theses and reports. These sources are mostly from the economic, management and social sciences, but also from the health and environmental sciences.

While no examples could be found of the use of the High Road/Low Road scenarios in the water sector, it is likely, judging from the fact that Sunter presented these scenarios to such a wide range of audiences, that some members of government and other stakeholders in the water sector would have been exposed to them in the late 1980s or early 1990s. South Africa's new water legislation certainly reflects the thinking associated with the High Road scenario, with emphasis on introducing ground-breaking new principles into the governance of South Africa's water resources. Though somewhat outdated now, the High Road/Low Road scenarios serve as a reminder of where South Africa could be headed at any point in history. In terms of water resources, South Africa is in need of thoughtful planning, development and management if its water resources are to continue to meet the needs of its ever growing and developing population.

Overview and Process

The Mont Fleur scenarios were developed in South Africa between 1990 and 1994. Key events during this time were the release of Nelson Mandela, and the legalisation of the ANC, Pan African Congress (PAC) and South African Communist Party (SACP) [36]. The country's first racially inclusive elections were also held at this time. Given this political climate, multiple forums emerged that brought a broad range of stakeholders together to try to develop a new way forward for South Africa. In particular, issues such as housing, education, and constitutional reform received attention [35, 36].

The Mont Fleur scenarios formed a part of this process and essentially tried to encourage debate, thinking and imaginative ideas around how to shape the first ten years of the 'new' South Africa and also to illustrate how certain choices would steer the country towards different outcomes. The Mont Fleur scenario team was made up of a diverse group of 22 prominent South Africans, including politicians, activists, academics and business people [36].

The Ostrich scenario represents a continuation of the status quo in South Africa and suggests that no negotiated settlement would be reached and that government would continue to be non-representative [37].

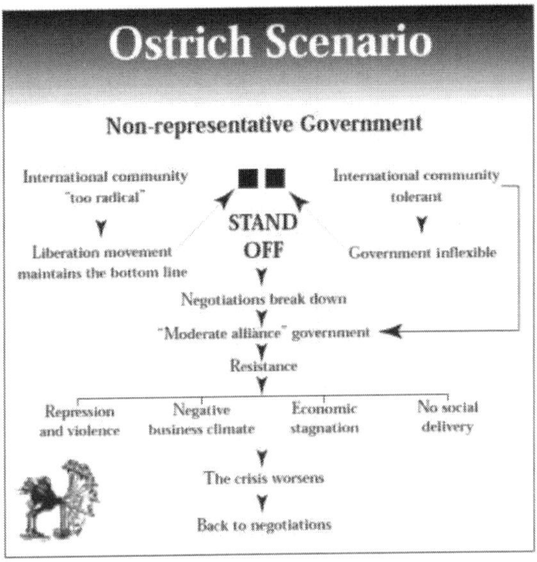

Figure 4. The Ostrich scenario [36]

The Lame Duck scenario suggests a South Africa where a settlement would have been achieved but where the transition to a new dispensation would be slow and indecisive [37].

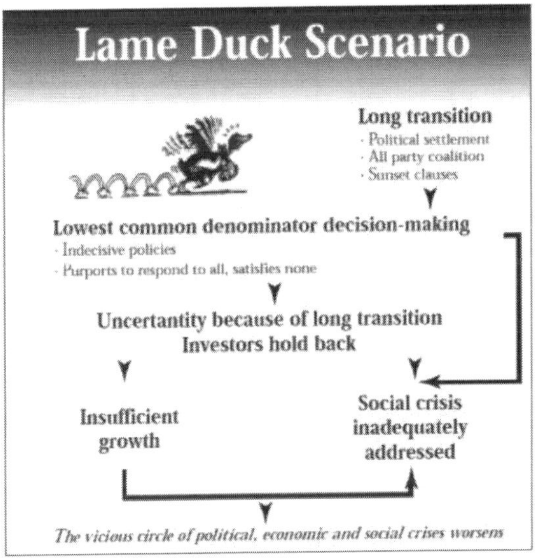

Figure 5. The Lame Duck scenario [36].

The Icarus scenario suggests a rapid transition to a new government that would push for populist and unsustainable economic policies [37].

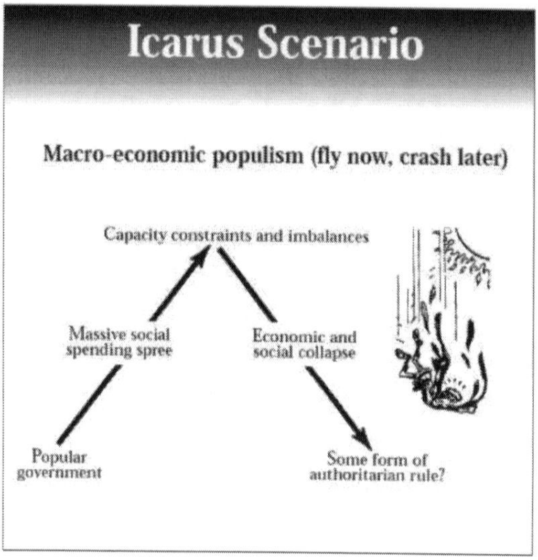

Figure 6. The Icarus scenario [36]

The Flight of the Flamingos scenario depicts a government that would choose sustainable policies that would lead the country towards inclusive growth and a maturing democracy [35, 36, 37].

Figure 7. The Flight of the Flamingos scenario [36]

By means of a process of negotiation and reflection on different drivers and concerns, the Mont Fleur scenario team was able to articulate a range of potential outcomes for South Africa during the 1992 to 2002 period. This also helped to clarify the goals and aspirations related to where the country should be heading.

Dissemination and Impact

A variety of dissemination techniques were used by the Mont Fleur team. Key to this process was the fact that each of the individual participants took responsibility for spreading the message of these scenarios. They did this by presenting and discussing the scenarios with more than 50 different groups of people including political parties, companies, academics, trade unions and civil society organisations [36]. This was possible given the diverse background that the team came from. Over and above this process, the scenarios were condensed into an easily accessible 14 page document. This document was distributed to national newspapers. A short video was also produced that combined cartoons with presentations by team members [35, 36].

The impacts of this scenario development process are subtle. Three key points are important in this regard. Firstly, Mont Fleur, along with other processes taking place in South Africa at the time, helped to establish a common language and understanding about the challenges facing the country and the way forward. This was because participants focused on an issue of common concern for all: 'the future of South Africa'. Secondly, although participants could not agree on one major solution to South Africa's problems, they could agree that certain solutions would not work (such as armed revolutions, continued minority rule and socialism). Thirdly, through an informal process of open conversation, participants who had not expected to agree with each other found common ground and shared understandings about the future of the country [36]. Given these points it is clear that the impact that the Mont Fleur scenarios had was first and foremost on the individuals who participated in the process. There was subsequently a more indirect impact on broader society once these individuals started presenting the scenarios to their various constituencies. Given the widely publicised nature of South Africa's political transition, these scenarios also gained popularity overseas [35].

The Mont Fleur scenarios have also been cited in a range of publications. These citations occur in journals, books, conference papers, dissertations and magazines that focus on a range of different disciplines, namely the social, natural and technical sciences. Given this broad interest, the publications cover a broad range of topics most of which are geared towards futures research, democratic transition and strategic planning. This citation record illustrates that the Mont Fleur scenarios seem to have had a considerable impact on the academic community.

Whilst the Mont Fleur scenarios are not obviously related to the South Africa water sector, they did contribute to setting a precedent for using scenario development for planning purposes in South Africa. So, for instance, as mentioned above, the National Water Resources Strategy established a set of water demand scenarios. As with the High Road/Low Road scenarios, the Mont Fleur scenarios were part of the thinking and move towards democratic transition in South Africa. As a result of and in order to complement this change, the water sector was fundamentally transformed and restructured.

Dinokeng Scenarios

Overview and Process

The Dinokeng scenario team consisted of 35 leaders from civil society, government, business, political parties, public administration, trade unions, religious groups, academia and the media. The scenario development process was sponsored by the financial institutions Old Mutual and Nedbank who believed that, 15 years into South Africa's democracy, it was important to initiate a reflective and constructive debate about the country's future. According to the Dinokeng scenario team, some of the most prominent challenges facing South Africa are unemployment and poverty, safety and security, education and health. These challenges appear all the more grave in the context of a volatile global economic market, and a global economic crisis that shook the world when these scenarios were developed in 2009 [30].

The Dinokeng scenario team agreed that South Africa needs to realise that the country has failed to appreciate or understand the imperatives of running a modern democratic state, and that there is a problem with the country's self-interested, unethical and unaccountable leadership across all sectors. Additional problems include a weak state that is increasingly less capable of addressing the country's critical challenges, and a population that is either not interested and is showing a growing dependence on the state to provide for everything, or has become co-opted into government or party structures since 1994 [30].

The scenario team developed three possible scenarios which the country could be heading into:

Firstly, the Walk Apart scenario suggests the state becoming increasingly weak and ineffective, and the population, which is looking out for its own interests, eventually losing patience with the state and resorting to protest and unrest to make its views heard. Because the state is unable to meet the population's demands and expectations, it responds brutally, and the result is a spiral of resistance and repression. The Walk Apart scenario therefore suggests a need for South Africans to address their critical challenges, to build state capacity and to organise themselves to engage government in a constructive way, in order to prevent themselves from heading towards disintegration and decline [30].

Secondly, the Walk Behind scenario suggests the state becoming increasingly confident and strong in terms of leading and directing development, fuelled by the fact that civil society is becoming more and more dependent and compliant. The problem is that the state does not have the capacity to address the critical challenges the country is facing on its own. The message of this scenario is that state-led development cannot be successful if there is insufficient state capacity. Furthermore, if the state intervenes constantly and dominates all other sectors, it will crowd out private business and civil society initiatives and will end up creating a population that is complacent and dependent on the state [30].

Thirdly, the Walk Together scenario suggests the state becoming collaborative and increasingly listening to its citizens and leaders from different sectors, engaging with critical voices, and consulting and sharing authority in order to work towards long-term sustainability. In this scenario there is also a focus on a population that takes leadership and holds government accountable and shows an active interest in policy development and outcomes. It is important that South Africans re-engage, that the capacity of the state is strengthened and that leaders from all sectors think beyond their own self-interest and contribute to nation-building [30].

In conclusion, the present contains the seeds for all three futures to be realised. For a healthy democracy and strong socio-economic development to persist, it is important to have a healthy interface between an effective state and an alert and involved population; the nature of this interface is likely to determine the future of the country [30].

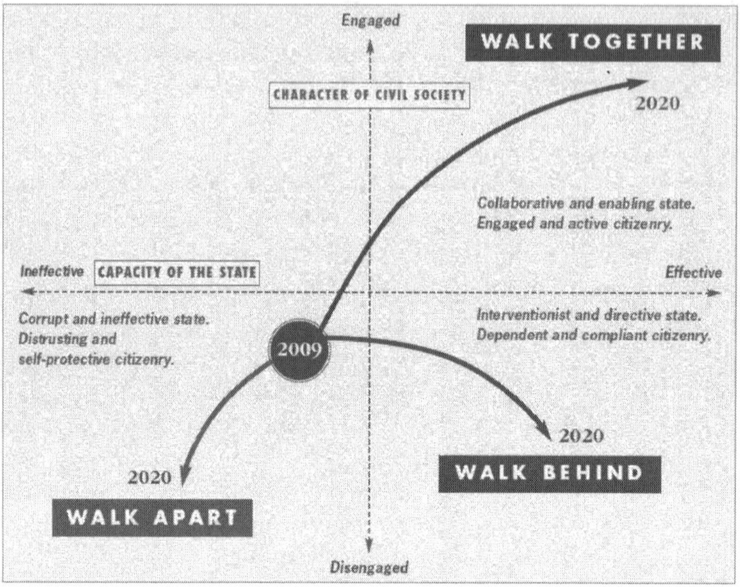

Figure 8. The Dinokeng scenarios [30]

Dissemination and Impact

In terms of dissemination and impact, once the Dinokeng scenarios on possible futures for South Africa had been developed, the messages of these scenarios were shared with a range of stakeholders. This engagement was followed up with a media and engagement campaign to communicate the Dinokeng scenarios to a variety of organisations, groups and communities across South Africa [30]. The Dinokeng scenarios and the process around their development were also placed on the Dinokeng scenarios website, which is a user-friendly resource for those who are interested in finding out more about these scenarios. The Dinokeng scenarios text is also available for download here.

A database search showed that these scenarios have been cited in a wide range of publications. These publications include discussion papers, theses, conference presentations, books and journal papers. The topics of the publications that cited the Dinokeng scenarios are wide-ranging and include issues around interrogating and addressing social issues related to South Africa's democracy, such as local government, education, housing, poverty, unemployment and food security. Many of these topics have a future-centred focus, e.g. investigating South Africans' perceptions about the future, or planning for the future in local government structures. The fact that the Dinokeng scenarios were cited in different kinds of publications and across different subject matters indicates that, at least among the research community, the scenarios were widely distributed and taken up by researchers from different social science-based backgrounds and interests.

A question that arises here is to what extent the Dinokeng scenarios may be of relevance to water resources planning, development and management. While no examples of their use in the water sector were found, it can be argued that the insights provided by these scenarios would prove valuable in focusing on resolving some of the water governance related issues South Africa is currently facing. Examples include problems around water pollution resulting from ineffective waste water treatment and mine and industrial effluents, and water service delivery to previously disadvantaged communities. Those who need to address these water governance related problems could benefit from taking into account the need for maintaining a balance between strong and effective leadership in all sectors and an interested and engaged population, and reflecting on the different future directions such a relationship or lack thereof could take.

Overview and Process

An example of scenario development with particular reference to the South African water sector is the South African Water Research Commission's (WRC) Water Sector Institutional Landscape by 2025 scenarios, developed by the authors in 2011 with the assistance of Chan tell Illbury as facilitator, and in consultation with a range of water sector related experts and stakeholders. The focus of the scenario development was on water resources management in South Africa, also with relevance to the water services sector. The aim of these scenarios was to build knowledge about key drivers and uncertainties that relate to the future of the South African water sector, and specifically about the context in which water institutions may operate in future [23].

The knowledge for this project was generated through a structured research process to target existing and new institutional structures and to ensure the involvement and participation of a broad range of stakeholders. The aim of this engagement was to identify water-related needs, priorities and uncertainties based on a wide range of perspectives. A broad range of methods was employed to include stakeholders from both rural and urban environments and with different cultures and educational backgrounds. These included interactive workshops, semi-structured interviews, and a web-based survey. This process was characterised by continuous assessment, learning and adaptation [23].

The key drivers and uncertainties that were identified were subsequently translated into different scenarios that hold potential implications for social and economic development, as well as water resources and services in South Africa. The four scenarios were derived from a matrix with two axes that represent the ability of the decision-making paradigm of water institutions to deal with 'complexity' (refer to the x-axis of the diagram), and the reconciliation of environmental, social and economic demands of present and future generations (referred to as 'sustainability' on the vertical or y-axis of the diagram) [23].

Four possible scenarios emerged from the matrix. The Greedy Jackal scenario depicts a South Africa where water is scarce but government still struggles to meet developmental demands and address backlogs. Under these urgent socio-economic circumstances, environmental responsibility is not prioritised. Despite this the need for a multidisciplinary response to complex water challenges is acknowledged [38].

The Wise Tortoise scenario suggests that a paradigm shift has occurred resulting in a water sector that is multi-layered and engages many different sectors given the strategic importance of the resource in all facets of development. This approach allows for proactive management rather than crisis response to challenges [38].
The Busy Bee scenario suggests that the water sector is defined by great intentions but does not follow up on these with necessary actions. Thus, whilst rhetoric embraces sustainability, in practice there is limited economic and social development to support this process. Part of the challenge is a lack of civil society engagement and failure to embrace the complexity facing water resources management [38].

The Ignorant Ostrich scenario suggests that government fails to recognise water as central to development. As such they rush to implement politically appealing but imbalanced and short term solutions. Civil society is not engaged in decision-making and the complexity inherent to the water sector is overlooked [38].

Figure 9. The Water sector institutional landscape by 2025 scenarios [38]

Dissemination and Impact

The scenarios were printed by the WRC in the form of a colourful booklet and subsequently have been disseminated to some stakeholders. The scenario document and technical report documenting the scenario development process are also available online. While much more could have be done in terms of dissemination, this was not a component required by the project's funders and was therefore not planned into the project process from the start of the project. Therefore no funding was available to carry out this important part of the scenario development process. Nonetheless, these scenarios have the potential of feeding into the decision-making processes of water resources managers and decision-makers, but could also potentially empower a range of other role players in the water sector to engage in participative governance [23].

By studying the dissemination, impact and lessons learned from the South African scenarios discussed above, along with other literature related to the impact of scientific research and the science-policy interface, it is possible to distil some lessons and challenges relating to impact and how to more effectively produce and disseminate impactful scenario products. A discussion on the impact of scenarios in general and reflections on such impact follows below.

The Impact of Scenarios

The previous section explored a number of South African scenarios in terms of their contents and impact. In terms of impact, Chantell Illbury and Clem Sunter refer to the "Wack" test, based on the ideas of Pierre Wack, a key scenarios planner in the 1970s and 1980s. According to this test, scenarios are not deemed important because of their prediction capability. What is important is their ability to influence the mindsets of decision-makers and to encourage them to act [39].

The issue of scenario impact is in many ways tied to a broader issue often referred to as the science-policy or science-end user interface. This issue essentially speaks to the challenge of getting knowledge that is produced by scientific or expert teams to be used in the public domain. This discourse recognises that there should be a close relationship between science or research products and their end-users, which could include government, policy-makers, businesses and communities. In reality, however, this relationship is not always an effective one, resulting in research often (or mostly) having minimal impact on policy and practice. The science-policy interface discourse explores why this happens in order to try to advise scientists and end-users about how to more effectively incorporate research into practice [40, 41, 42].

In terms of scenarios there tend to be two major opportunities for impact. The first is an impact on the participants who are part of the scenario development process. This is referred to as 'communication forscenarios' [43]. Similarly this opportunity for impact can be referred to as first order influence. First order influence refers to participants in the scenario development process undergoing personal changes in their thinking and behaviour. They also commit to the process, learn new skills, and build new networks and relationships. Because participants increasingly respect, understand and trust each other, they jointly commit to change [37].

The second is the impact of scenarios on broader society. This can be referred to as 'communication ofscenarios' [43]. Here a wider group of stakeholders ideally need to be exposed to the scenarios once they are fully developed. As such, at this stage it is important to think about ways to foster appropriate dissemination and use of scenarios. This stage can also be referred to as second and third order influence. Second order influence is closely linked to first order influence. Participants who have been part of the scenario development process go back to their communities and networks and start sharing their new language, thoughts and insights with others. Third order influence is a process of social change, but can be difficult to monitor and study because of the many variable factors that influence every change process [37].

The following sub-sections reflect on the impact of scenario development on the participating team as well as the impact or influence of scenarios on broader society.

Impacts on Participants in the Scenario Building Process

Participants in a scenario development process actively engage and transform the process in the sense that they are asked to share their views, ideas, concerns and experiences in order to generate drivers to develop scenarios or stories from these drivers. It is important to recognise that this kind of individual impact is difficult to quantify and tends to be very subtle [35; 37]. Nonetheless, the kinds of impacts that individuals experience can include:

- Experiencing reframed mental models – By being forced, through the scenario development process, to articulate and share different perspectives and mental models, participants are made to think carefully about their perceptions and often re-think their views when faced with other participants' views and the need to move collectively towards a desired future [44].
- Gaining a broadened network of relationships – Scenario development processes bring together groups of people to have open and constructive conversations. This process fosters a shared understanding, trust and a sense of community [44].
- Regenerating energy, commitment, and action – By clarifying desired futures and building consensus about how different actions will navigate society towards certain scenarios, a sense of regenerated energy and commitment can be achieved. Also, with new commitment in place, new actions can be catalysed [44].
- Taking pride in participation – When interviewed, participants tend to be quite proud of their involvement in scenario development processes. This encourages them to use and share the learning from the scenario development process during other projects and/or engagements [35].
- Creating a common vocabulary, trust and mutual understanding – Through the process of developing scenarios these subtle processes tend to be fostered. This is important as it is through trust and understanding that people are able to work together towards a desired shared future [37].

Facilitating and Forming a Scenario Team

Whilst it is clear that a subtle process of impact and transformation can occur in a scenario team, this does not happen automatically. There are a number of lessons that have been learned through scenario development processes over the years that need to be borne in mind.

Firstly, having a diverse team is important [45]. The team should come from different age, race and gender brackets as well as a wide range of ideological spectrums [35]. This diversity is important because the more diverse the team is, the more diverse the driver inputs will be and as such the richer and more accurate the scenario development process will be. Also, an inclusive rather than exclusive scenario development process lends legitimacy to the process [46].

Secondly, embracing transdisciplinarity in any scenario development process is important. This implies that in order for scenarios to have the impact they need, they should be produced by a team made up of multiple different actors from government, civil society, communities, and research institutions. This will help the team to take into account different types of knowledge that different actors have (such as technical, traditional, experiential, cultural, and political knowledge). In so doing the inherent complexity in future planning processes will be reflected [47].

Finally, working with a diverse team with different knowledge, experience and viewpoints is not always easy. Conflict can arise when participants with different viewpoints are made to work together. Also, meaningfully incorporating feedback from diverse sets of stakeholders tends to be a highly time consuming process. Given these and other challenges that can arise, the importance of having a skilled, sensitive and insightful facilitator cannot be underestimated. Such a facilitator needs to be able to manage strong individuals who dominate conversation with their own agendas, and needs to be able to encourage everyone to express their opinions during the scenario development process [35].

The impacts of scenarios on broader society are harder to ascertain and measure than the impacts of scenario development on the scenario team itself. This is because there are no measurable criteria for quantifying the impact that scenario products have on society, be they in written or oral form. Also the outcome of scenario development processes can never be attributed to a single factor. Scenario development processes typically deal with broad developmental issues making the range of issues and actors that they try to affect diverse. Scenario development processes also happen within the context of a range of related social activities, such as developments in policy, civil society events and public debates. For example, in the case of developing the South African scenarios of the 1980s and 1990s, there were multiple social forums, political parties, and government groups working on transforming the country. These scenarios and their related processes were just one input amongst many others that were part of the broad transition process. Similarly, the Water Sector Institutional Landscape by 2025 scenarios exist alongside scenarios established by the National Water Resources Strategy, the various government departments that do strategic planning and forecasts in relation to water, and the host of grassroots organisations that work on managing water sustainably for the future. Any impact or change in the water sector must then be attributed to a whole range of interlocking factors rather than just one set of scenarios.

Facilitating the Effective Dissemination of Scenario Products to Society

In order for scenarios to have influence in the broader public space a number of key lessons are important. Firstly, a broad and extensive communication process is a key requirement and should be planned and budgeted for from the beginning of the project [42]. It is important that such a process targets multiple different actors in society, and takes place at many levels of scale (local, provincial, national) [23, 41, 42] in order to engage society and attempt to create a better future [37]. Non-government actors are an important target audience because they are critical in terms of instigating social debate, bringing about grassroots changes and challenging authorities to improve their performance [42].

In government, actors need to be aware of scenario products and how they can make use of them [23]. With regard to the South African water sector in particular, there seems to be a need to enable officials from DWA to apply the outcome of scenarios thinking and processes in their strategic decision-making aimed at mapping out the future of the water sector. A possible way of enabling experts and government officials to think imaginatively and creatively about the future, given their considerable daily workload and challenges, would be to involve scenario experts as facilitators for strategic planning sessions. Such sessions should ideally take the form of one or two day workshops in order to remove government officials from their immediate working environment and enable them to apply their minds to thinking creatively and focusing exclusively on the planning task at hand [23]. When engaging with government departments, it is important to be sensitive to and aware of different issues inherent in the government hierarchy. Non-political, technical experts tend to have a good knowledge of technical issues, but it is also important to target more senior political actors as they tend to have more decision-making power and can therefore implement changes and ideas brought about through the scenario development process more effectively [42].

It has been argued that regardless of which actor is being focused on, there are three key points to bear in mind in terms of targeting actors with information. A clear plan of action needs to be laid out and followed up on. The information needs to be shared in a manner that is non-threatening, interactive and flexible. Scenarios can be disseminated by tapping into existing networks and events such as management meetings, seminars and the media [42].

In addition, the way that scenarios are packaged and communicated is important [42]. There is a whole host of ways that information can be packaged and disseminated. There can be face-to-face dissemination [23], where scenarios are verbally presented at workshops, conferences, public gatherings, business breakfasts, and corporate events. Style of presentation is crucial in this regard. The presentations need to be simple, clear and memorable. The presenter needs to be engaging and open to feedback from the audience [35]. Radio or television documentaries can also be utilised to disseminate scenario ideas and generate public debate [36, 40, 42].

Another option is to publish the scenarios in a written format. A range of media can be used. The scenarios can be published in books, illustrated pamphlets [23] and newspapers [48]. Cartoon artists can be brought on board to illustrate the scenarios. Magazines and web pages can also be targeted. Written documentation about scenarios has proved to be a successful model. For example, Sunter and Illbury's 'The Mind of the Fox: Scenario Planning in Action' [49] is popular reading material and widely distributed.

Finally, it is crucial that the scenario products are seen as legitimate from the start of the scenario development process. They need to have buy-in from influential people involved in the issue that the scenarios explore. This legitimacy is generated by ensuring that the facilitators of the process as well as the scenario team are respected. Although a range of actors must be included in the scenario development process, and must be targeted in the dissemination process, it remains important to include high level and well-connected people in the team as it is often these individuals who will provide the 'insider' links for scenarios to be heard and disseminated through channels of influence [42]. If these strategic individuals cannot be made part of the team, they need to be made aware of and kept informed about the scenario development process to secure their interest and support [35].

Dissemination is not without its challenges. It is challenging to disseminate in a way that suits and reaches a diverse audience with different languages, levels of education, varied professional backgrounds and cultures. Another challenge of ensuring the uptake of scenarios (and research in general) is that dissemination is often not part of the project planning process, and as a result funding often runs out before scenario uptake and use can be promoted [48]. Also, depending on how it is done, the dissemination of scenarios can be very expensive [35].

Over and above the specific processes linked to the impact of scenarios on the scenario team and broader society there are some general points that are important to bear in mind when planning for impact in relation to scenario products.

Firstly, when starting the scenario development process, it is important to be clear about the purpose of the process one is undertaking and designing it accordingly. What are the intended outcomes of the process? Who is the process meant to influence and what product(s) will be necessary for this to happen [35]? Essentially scenarios need to fill a strategic gap or opportunity in society [50].

Secondly, questions also need to be asked about the timing of the scenario development process. Is there likely to be sufficient recognition among the intended target audience(s) that the problem being addressed is important and that the process is therefore potentially beneficial? Is the political environment such that intended target audience(s) will be responsive to fresh, unorthodox thinking [35,42]?

Thirdly, attention also needs to be paid to the legitimacy of those financing and promoting the process, and the credibility of the project team developing the scenarios in the eyes of both the sponsors and the target audience(s) [35, 40, 42].

CONCLUSIONS

In conclusion, it seems that since the initial High Road/Low Road scenarios were developed, scenario development has taken root in South Africa, with several follow up scenarios having been developed since [28]. This development suggests that South African decision-makers must deem scenario development to be of considerable importance and utility, as it is often government or government-related institutions that develop or commission new sets of scenarios. These subsequent scenarios seem to mirror their predecessors with their snappy titles and straightforward structure and certainly have the potential to inspire decision-makers with regard to their planning activities [28].

Based on the discussion and reflections above, scenario development should involve a focus on dissemination and impact from the onset of the scenario process. Impact can happen at the level of participants in the scenario development process as they are exposed to new ideas and start adopting a new way of thinking about current issues of importance. These ideas have the potential to slowly infiltrate the networks of these participants and to also influence their thinking. At the same time, it is important to have a strong dissemination process in order to reach as many people as possible beyond the project team. The High Road/Low Road presentations are an example of a highly effective dissemination process made possible by an engaging speaker and interesting topic that was clearly and simply brought across to a wide range of audiences. Another key method of dissemination is to raise awareness about where the scenarios can be found and to make it easy for people to access them. The open access route followed by the Dinokeng scenario team is a good example of a scenario document that is easily available on a website, accompanied by much useful background information. It is this dissemination phase that has been lacking in the Water Sector Institutional Landscape by 2025 scenarios, and a follow up process is needed to plan how more people could be made aware of these scenarios and their usefulness to decision-making and planning in the South African water sector.

It is also important to keep in mind that scenarios are likely to have a higher impact if they are developed with the intention of identifying or solving particular problems [51]. If there is an intended target audience with particular information needs from the beginning of the scenario process, the scenario team will be able to keep this in mind when developing the scenarios. This will also ensure more effective uptake of the scenarios as pre-defined end-users exist. In the water sector, for instance, it could be effective for decision-makers who are grappling with a particular issue to solicit scenario inputs to aid them in making decisions regarding that issue.

In terms of future research, three areas come to mind based on what has been discussed in this chapter:

Firstly, a large scale study (mostly comprising of interviews) is needed to understand in greater detail the impact of scenarios on scenario participants, society and government planning processes [28]. Much of what has been argued to date in terms of the impact of scenarios has been on the basis of inference and assumptions. It would be interesting, though admittedly also very difficult, to try to substantiate views around the impact of scenarios with empirical evidence.

Secondly, it would be important to study how a scenario team would know that the timing is right to come up with and disseminate a new set of scenarios. It is reasonably easy to see that scenarios would have been important for particular moments in history, for example the political transition in South Africa, but it is considerably more difficult to determine when there may be an ideal window of opportunity in future in which scenarios may make an impact. It may also be important to determine which factors other than and in support of ideal timing would be important for scenarios to achieve impact.

Thirdly, building on this chapter, it would be important to determine how best to ensure that scenarios can become more useful and practical to policy-makers and other end-users. How can scenario teams ensure that end-users know how best they may use scenarios in order to influence their future planning? The issue of providing navigation to and between different scenarios and future outcomes is important in this regard.

Clearly, scenario development is a useful process to help decision-makers cope with and plan amidst uncertainty. Particularly in the context of the South African water sector, it is important to recognise that uncertainty is deepening in many ways given the impending presence of multiple stressors such as climate change, basin closure, growing populations, migration movements and a growing economy. These stressors, along with the institutional fluctuations and changes within the water sector itself, make it increasingly important for decision-makers to work with scenarios to help them to plan sensibly and creatively despite uncertainty. However, in order for scenarios to be useful it is important to plan for and carefully think about how to maximise their impact.

ACKNOWLEDGEMENTS

The authors would like to acknowledge the CSIR's librarian, Engela van Heerden, for her excellent work in sourcing a large amount of relevant literature that contributed to this chapter. They would also like to acknowledge Wilma Strydom for her valuable review comments on this chapter. Finally, the authors would like to acknowledge the Water

Research Commission (WRC), the organisation that funded the development of the Water Sector Institutional Landscape by 2025 scenarios. The learning that the project team gained from this project was a key input into this chapter.

REFERENCES

1. Bogardi JJ, Dudgeon D, Lawford R, Flinkerbusch E, Meyn A, Pahl-Wostl C, Vielhauuer K and Vörösmarty C. Water Security for a Planet Under Pressure: Interconnected Challenges of a Changing World Call for Sustainable Solutions. Current Opinion in Environmental Sustainability 2012; 4 35–43.
2. Slaughter RA.Welcome to the Anthropocene. Futures 2012; 44 119–126.
3. WWAP (World Water Assessment Programme). The United Nations World Water Development Report 4: Managing Water under Uncertainty and Risk. Paris: UNESCO; 2012.
4. UN (United Nations). Declaration of the United Nations Conference on the Human Environment. Stockholm, 5-16 June 1972.
5. Brundtland GH. Address by Mrs Gro Harlem Brundtland, Chairman at the Closing Ceremony of the Eighth and Final Meeting of the World Commission on Environment and Development 27 February 1987. Tokyo, Japan.
6. WMO (World Meteorological Organisation). International Conference on Water and the Environment: Development Issues for the 21st Century. Geneva: ICWE Secretariat; 1992.
7. UN (United Nations). Agenda 21. United Nations Conference on Environment and Development Rio de Janeiro, Brazil, 3 to 14 June 1992.
8. Spangenberg JH, Pfahl S, Deller K. Towards Indicators for Institutional Sustainability: Lessons From an Analysis of Agenda 21. Ecological Indicators 2002; 2 61–77.
9. UN (United Nations). World Development Report, 1980. Washington, D.C.: The World Bank; 1980.
10. UN (United Nations). World Development Report, 1985. New York: Oxford University Press; 1985.
11. UN (United Nations). World Development Report, 1990: Poverty. New York: Oxford University Press; 1985.
12. UN (United Nations). World Development Report, 1995: Workers in an Integrating World. New York: Oxford University Press; 1995.
13. World Bank. Indicators. http://data.worldbank.org/indicator. (accessed 30 July 2012).

14. Bates C, Kundzewicz ZW, Wu S and Palutikof JP., editors. Climate Change and Water. Technical Paper of the Intergovernmental Panel on Climate Change. Geneva: IPCC Secretariat; 2008.
15. Davis CL. Climate Risk and Vulnerability: A Handbook for Southern Africa. Pretoria: Council for Scientific and Industrial Research; 2011.
16. Tewari DD. A Detailed Analysis of the Evolution of Water Rights in South Africa: An Account of Three and a Half Centuries from 1652 AD to the Present. Water SA 2009; 35(5) 693-710.
17. Gleick PH. The World's Water, 1998-1999. Washington DC: Island Press; 1998.
18. DWAF (Department of Water Affairs and Forestry). National Water Act of South Africa. Act No. 36 of 1998. Pretoria: Department of Water Affairs and Forestry; 1998.
19. DWAF (Department of Water Affairs and Forestry). National Water Resources Strategy – First Edition. Pretoria: Department of Water Affairs and Forestry; 2004.
20. DWA (Department of Water Affairs). Annual Report of the Department of Water Affairs Vote 37: 1 April 2010 To 31 March 2011. http://www.pmg.org.za/minutes/30. (accessed 6 August 2012).
21. Emerson JW, Hsu A, Levy MA, de Sherbinin A, Mara V, Esty DC, Jaiteh M. Environmental Performance Index and Pilot Trend Environmental Performance Index. New Haven: Yale Center for Environmental Law and Policy; 2012.
22. Green GC, Day JA, Mitchell SA, Palmer C, Laker MC, Buckley CA. Water Research Commission 40-Year Celebration Conference - Syntheses of Themed Sessions. Water SA 2011; 37(5) 609-618.
23. Claassen M, Funke N, Nienaber S. 2011. The Water Sector Institutional Landscape by 2025 Technical Report. WRC Report No. 1841/1/11. Pretoria: Water Research Commission; 2011.
24. Mietzner D, Reger G. EU-US Seminar: New Technology Foresight, Forecasting and Assessment Methods. Seville; 13-14 May 2004.
25. Landeta J. Current Validity of the Delphi Method in Social Sciences. Technological Forecasting and Social Change 2006; 73 467-482.
26. Daum J. How Scenario Planning Can Significantly Reduce Strategic Risks and Boost Value in the Innovation Chain. The New Economy Analyst Report 8 September 2001.
27. Sunter C. The World and South Africa in the 1990s. Cape Town: Human Rousseau Tafelberg; 1987.
28. Galer G. Scenarios of Change in South Africa. The Round Table 2004; 93(375) 369-383.

29. DACST (Department of Arts, Culture, Science and Technology). South African National Research and Technology Foresight Project. Pretoria: DACST; 1999.
30. Dinokeng team. 3 Futures for South Africa. www.dinokengscenarios.co.za (accessed 10 August 2012).
31. (WBCSD) World Business Council for Sustainable Development. Business in the World of Water: WBCSD Water Scenarios to 2025. www.wbcsd.org/web/H2OScenarios.htm (accessed 30 July 2012).
32. GRA (Global Research Alliance) Science and Technology-Based Water Scenarios for Sub-Saharan Africa. www.research-alliance.net. (accessed 6 August 2012).
33. Peterson GD, Cumming GS, Carpenter SR. Scenario Planning: A Tool for Conservation in an Uncertain World. Conservation Biology 2003; 17(2) 358-366.
34. Finlev T. Future Peace: Breaking Cycles of Violence through Futures Thinking. Journal of Futures Studies 2012; 16(3) 47-62.
35. Segal N. Breaking the Mould: The Role of Scenarios in Shaping South Africa's Future. Stellenbosch: SUN Press; 2007.
36. Kahane A. The Mont Fleur Scenarios: What Will South Africe Be Like in the Year 2002? Deeper News 1992, 7(1) 1-22.
37. Maddison S, Cronin D, Williams S, Coggan R. Democratic Dialogue: Finding the Right Model for Australia. Indigenous Policy and Research Unit, Discussion Paper No.1. Sydney: University of New South Wales; 2009.
38. Claassen M, Funke N, Nienaber S. The Water Sector Institutional Landscape by 2025. WRC Report No. TT 514/11. Pretoria: Water Research Commission; 2011.
39. Karuri-Sebina G, Rosenzweig L. A Case Study on Localising Foresight in South Africa: Using Foresight in the Context of Local Government Participatory Planning. Foresight 2007, 14(1) 26-40.
40. Strydom WF, Funke N, Nienaber S, Nortje K, Steyn M. Evidence-based Policy-making: A Review. South African Journal of Science 2010; 106(5/6) 249 – 334.
41. Funke N, Nienaber S, Henwood R. Scientists as Lobbyists? How Science Can Make its Voice Heard in the South Africa Policy-Making Arena. International Journal of Public Affairs 2011; 10(102) 421 – 443.
42. Funke N, Nienaber S. Promoting Uptake and Use of Conservation Science in South Africa by Government. Water SA 2012; 38(1) 105-114.
43. Lebel L, Thongbai P, Kok K, Agard JBR, Bennett EM, Biggs R, Ferreira M, Filer C, Gokhale Y, Mala W, Rumsey C, Velarde SJ, Zurek M, Blanco H, Lynam T, Tianxiang Y. Sub-global Scenarios. In: Capistrano D, Lee M, Raudsepp-Hearne C, Samper C. (eds.) Ecosystems and Human Wellbeing:

Multi-scale Assessments. Findings of the Subglobal Assessments Working Group of the Millennium Ecosystem Assessment. Washington D.C.: Island Press; 2005. p229-259.
44. Ringland G. Scenario Planning, 2nd edition. England: John Wiley & Sons; 2006.
45. Audouin M, Preiser R, Nienaber S, Downsborough L, Lanz J, Mavengahama S. Exploring the Logic of Complexity for Researching Social-Ecological Systems. Ecology and Society In Press.
46. Ogilvy JA. Creating Better Futures: Scenario Planning as a Tool for a Better Tomorrow. New York: Oxford University Press; 2002.
47. Jacobs I, Nienaber S. Water Without Borders: Transboundary Water Governance and the Role of the 'Transdisciplinary Individual' in Southern Africa. Water SA (WRC 40 Year Celebration Special Edition 2011; 37(5) 665 – 678.
48. Communication with Clem Sunter. 26 July 2012. Pretoria.
49. Illbury C, Sunter C. The Mind of a Fox. Cape Town: Human & Rousseau/Tafelberg; 2001.
50. Johnson G, Scholes K, Whittington R. Exploring Corporate Strategy: Text and Cases. New Jersey: Prentice Hall, Financial Times Press; 2008.
51. Bohenksy E, Reyers B, Van Jaarsveld AS. Future Ecosystem Services in a Southern African River Basin: a Scenario Planning Approach to Uncertainty. Conservation Biology 2006; 20(4) 1051-1061.

CITATION

Nikki Funke, Marius Claassen and Shanna Nienaber (2013). Development and Uptake of Scenarios to Support Water Resources Planning, Development and Management – Examples from South Africa, Water Resources Planning, Development and Management, Prof. Ralph Wurbs (Ed.), ISBN: 978-953-51-1092-7, InTech, DOI: 10.5772/52577. Available from: http://www.intechopen.com/books/water-resources-planning-development-and-management/development-and-uptake-of-scenarios-to-support-water-resources-planning-development-and-management-e

Index

I

M

O

P

R

S

T

U

V

W

Z